Sarah Hussein

A proposed Web Indexing Model Based on Metadata

Sarah Hussein

A proposed Web Indexing Model Based on Metadata

Noor Publishing

Imprint

Cover image: www.ingimage.com

Publisher:
Noor Publishing
is a trademark of
International Book Market Service Ltd., member of OmniScriptum Publishing Group
17 Meldrum Street, Beau Bassin 71504, Mauritius
Printed at: see last page
ISBN: 978-620-2-78882-3

Ministry of Higher Education and
Scientific Research-Iraq
University of Technology
Department of Computer Science

A proposed Web Indexing Model Based on Metadata

Sarah Hussein Toman

سورة طه :114

Dedication

My Mother, Her
Shadow and Reflection.

Sarah....2014

Acknowledgment

Thanks and praise are to Allah, and prayer and peace be upon his messenger Mohammed, and his pure relatives and best companions. The research work reported in this thesis would not have been possible without the generous help of many persons, to whom I am grateful and wish to express my gratitude:

- I wish to thank my supervisor Dr. Rehab F. Hassan, for her guidance and encouragement. I would also like to thank her for having allowed me to work in the Web field.

- I would like to thank Dr. Asraa T. AL.Attar, for her assistance.

- Special thanks to my parents, my family, for their love and support to make my life sunny and colorful.

Sarah......2014

Contents

Chapter Six: Conclusions and Future Works

List of Abbreviations

Abbreviation	Description
APIs	Application Programming Interfaces
ARFF	Attribute-Relation File Format
ASCII	American Standard Code for Information Interchange
ASP	Active Server Page
CE	Content Extract
CPU	Central Processing Unit
DCMES	Dublin Core Metadata Element Set
DCMI	Dublin Core Metadata Initiative
DNS	Domain Name System
DOM	Document Object Model
ExBSN	Extraction Based on Sibling Nodes
ExBST	Extraction Based on Sub-Tree
GIF	Graphics Interchange Format
HTML	Hyper Text Markup Language
HTTP	Hyper Text Transfer Protocol
IE	Internet Explorer
IETF	Internet Engineering Task Force
IIS	Internet Information Server
I/O	Input/ Output
IP	Internet Protocol
IPv4	Internet Protocol version 4
IPv6	Internet Protocol version 6
IR	Information Retrieval
ISO	International Organization for Standardization
ISP	Internet Service Provider

Abbreviation	Description
JPEG	Joint Photo Graphic Experts Group
MP3	Motion Picture Experts Group Layer 3
MP4	Motion Picture Experts Group Layer 4
MWICD	Multilateral Web Indexing Client Dispatcher
MWIM	Multilateral Web Indexing Model
PDF	Portable Document Format
RDF	Resource Description Framework
SDK	Software Development Kit
SE	Search Engine
SEA	Search Engine Agent
SMTP	Simple Mail Transfer Protocol
SST	Site Style Tree
TCP	Transmission Control Protocol
TIFF	Tagged Image File Format
TPL	Task Parallel Library
UDP	User Datagram Protocol
URI	Uniform Resource Identifier
URL	Uniform Resource Locator
URN	Uniform Resource Name
UTC	Coordinated Universal Time
UTF-8	(Universal Character Set) Transformation Format 8-bit
W3C	World Wide Web Consortium
WWW	World Wide Web
XHML	Extensible Hypertext Markup Language
XML	eXtanded Markup Language
XPath	XML Path Language

List of Figures

List of Tables

ABSTRACT

The World Wide Web contains large volumes of Web pages that are poorly characterized in terms of both metadata and structure descriptions. Metadata is the data that describe data, for Web pages, it may provide a significant amount of information about the Web without examining its content and serves a variety of purposes.

At the other side, the search engine needs to scrape all Web pages of the target Website, when a new Web page is first available, it will typically take several days—for the page to be detected and indexed by search engine. Until a page has been indexed, the users cannot find information about it on search engines results. This represents a time and resource-consuming operation and heavy network bandwidth consumption involved with the retrieval of the mark-up code of these Web pages which finally affects the users by increasing the time taken for a new content or a content modification to be reflected in the search engines results. For instance, Web related metadata can improve search efficiency and the quality of the search results.

This thesis proposes an approach called **Multilateral Web Indexing Model (MWIM)**, which aims to describe a solution to the previously mentioned issues by establishing a more tied collaboration between Web sites and search engines. The major idea of this approach is exposing to the search engines an auto generated metadata about the structure of a Web site and the contents of each Web page. Two kinds of metadata have been proposed: **Sitemap** and **Content Extract** file. The Sitemap provides information about the structure of the Website and the Content Extract contains pre-processed information describing the contents of a Web page. The Document Object Module (DOM) tree has been used as an efficient tool for representing the Web page content, while the XML Path Language

(Xpath) has been used to provide a powerful syntax to address specific elements of DOM tree and extract metadata from HTML Web page.

Consequently, the search engine does not need anymore to retrieve the Web page mark-up code and perform the data extraction because this stage is performed directly on the Web site's platform when a Web page was created or modified.

The work of a search engine would be made a lot easier, less time consuming, resources and bandwidth, where the percentage of bandwidth savings is **72.9%**, i.e. the average size of a Content Extract metadata file being of one-third of the original HTML markup size. Thus, the search engine servers require less computational power to retrieve the forward index of a Web page.

CHAPTER ONE

INTRODUCTION

1.1 Introduction

The World Wide Web (commonly known as the web or WWW) is the biggest and most widely known information source that is easily accessible and searchable. It consists of billions of interconnected documents (called Web pages) which are authored by millions of people. It is an Internet-based computer network that allows users of one computer to access information stored on another through the world wide network called the Internet. The web would not be possible without the Internet, which provides the communication network for the Web to function [1]. One of the most common reasons for users to be hanging around on the internet every day is the abundance of information it is loaded with [2]. The information is generated by multiple sources and is carefully organized in the form of files and web pages, which, when grouped to form a single entity, become a website.

A website represents a centrally managed group of web pages, containing text, images, and all types of multi-media files presented to the attention of the Internet users in an aesthetic and easily accessible way.

All websites enabled through the Internet constitute the World Wide Web (WWW) [3].

With information being shared worldwide, there was a need for individuals to find information in an orderly and efficient manner. Thus the development of search engines began [1].

Search engines are special sites on the web that are designed to help people to find information stored on other sites. It works by storing information about a large number of web pages, which they retrieve from the WWW itself. The contents of each page are then analyzed to determine how it should be indexed [2].

The appearance of any website on the search engine result pages is also related to some kind of indexing. As soon as one's website is noticed by the search engine crawler, its contents are scanned and entered into their database of already scanned sites i.e. their index.

When a user types certain keywords in the search engine bar, the search results will be readily presented to him, since the search crawler has already scanned the documents and listed them in the order of their keyword relevance.

1.2 Motivation and Problem Domain

In an era where the WWW became a media which comprises a huge extent of the humanity's knowledge, news and a significant amount of daily activities such as socializing, or commerce, a user will most likely enlist for the help of a web search engine in a pursue to fulfill his/her quest with the results scattered in the tens of millions of web pages found in the World Wide Web.

Due to the original design of the WWW standards which does not provide any support for structured information, neither defines any convention to delimitate the main content of a web page from the aside contents such as navigation menus and the lack of a widely adopted conventions related to the mode how the relevant data should be exposed, a search engine needs to crawl a website and extract the data from the mark-up code of the web pages.

Also the lack of any record describing the modifications occurred to a website, imposes a need of periodically crawling each website found in WWW, to determine its structure, respectively to index the contents of its web pages to update the search engine's indexes in the case of any potential modifications of its contents. This represents a time and resource consuming operation which finally affects the users by increasing the time taken for a new content or a content modification to reflect in the search engines results. Also, the servers which host search engines require powerful hardware and expensive internet connections to make their indexer always up to date.

So, it will be so powerful and useful if all websites contain by themselves their own metadata files that are accessible by any search engine where these metadata files are extracted and available for any search engine; i.e. they do not need to be extracted by each search engine.

1.3 The Aim of Thesis

This thesis aims to propose a different approach for improving web indexing quality and lowering its workload through website-search engine coactions.

The proposed solution is achieved by constructing new structured information for the web-related metadata and new extension of the XML sitemap. By making the website have the ability to auto-generating metadata that describe the website contents and structure that are needed by the search engine to index the website in its database, which gives the website a faster reflection in the search results.

1.4 Contribution of the Thesis

The XML Sitemap and Meta Tag are Kinds of collaboration between website and search engine. In this thesis, new structured information of web-related metadata (Content Extract File (.ce)) is proposed with an extension of XML sitemap that creates a new collaboration between them.

This collaboration represents a different web indexing model achieved by using traditional methods. This model improves the search engine performance by making:

- The crawler with less time in crawling or re-crawling the website.
- The indexer does not need to parse the pages anymore, but just retrieve the relevant data from the website by loading Content Extraction Files (.ce), where these files are forward-index; i.e. contain data that can be directly indexed for all search engines without any other processing.

1.5 Literature Survey

The most important researches that have relation to the thesis research subject are:

1) Estievenart F. and et al, [2003] [3]: present "A tool-supported method to extract data and schema from web sites", that is, to extract the page contents as XML documents structured. All the pages are analyzed to derive their common structure. This structure is formalized by an XML document, called META, which is then used to extract an XML document that contains the data of the pages and the XML Schema validating these data. The META document can describe various structures such as alternative layout and data structure for the same concept, structure multiplicity, and separation between layout and informational content. XML Schemas extracted from different page types are integrated and conceptualized into a unique schema describing the domain covered by the whole website. This conceptual schema is used to build the database of a renovated website.

2) LAN YI [2004] [4]: Presents “ Web Page Cleaning for Web Mining”, This thesis focuses on the problem of Web page cleaning; i.e. the pre-processing of Web pages to automatically detect and eliminate noises for Web mining. Based on the DOM tree model, two novel Web page cleaning methods, i.e., the Site Style Tree (SST) based method and the features weighting method, are devised. Both methods are based on the observation that: in a given Website, noisy blocks of a Web page usually share some common contents and/or presentation styles, while the main content blocks of the page are often diverse in their actual contents and presentation styles.

3) Migletz J. [2008] [5]: Presents “Automated Metadata Extraction”, this thesis presents a new technique for automatically extracting metadata from files (.doc, .docx, .odt, .pdf, .mp3, .mp4, .jpeg, tiff, and .gif). The technique is embodied in a program called fiwalk that has a plug-in architecture allowing new metadata extractors to be readily incorporated. Output from fiwalk can be provided in multiple formats such as ARFF and text.

4) AL-Attar I. T. [2009] [6]: Presents “Design and Implementation of Web Search Engine”, in this thesis, she has built a search engine by crawling system with three techniques, enhanced indexing system by using a new stemming algorithm and extracting additional information and retrieve the most relevant web pages with the queries that are presented from the users by using an improved web page ranking system. The conclusions of this thesis are reducing the time and storage space of web pages downloading, reducing the number of iterations which yields to reduce the spent processing time.

5) Yang J. and et al [2009] [8]: Present "Incorporating Site-Level Knowledge to Extract Structured Data from Web Forums", this paper studies the problem of structured data extraction such as post title, post author, post time, and post content from various web forum sites. They incorporate both page-level and site-level knowledge and employ Markov Logic Networks (MLNs) to effectively integrate all useful evidences by learning their importance automatically.

6) Choochaiwattana W. [2012] [7]: Presents "An Algorithm of Product Information Extraction from Web Pages: a Document Object Model Analysis Approach", this paper improves the efficiency of the product search engines by embedded classification and extraction features in web crawlers. It focuses only on the extraction features and investigates how well the DOM tree structures of web pages contributes to product information extraction task when they are used to identify pieces of the product information on web pages. Two algorithms for product information extraction tasks have been proposed: Extraction Based on Sub-Tree (ExBST) and Extraction Based on Sibling Nodes (ExBSN).

1.6 Thesis Outline

This thesis is presented in five chapters; in addition to the present one:

Chapter Two: gives an overview of Basic WWW Technologies (Web Browser, Web Server, Protocols, Addressing, and Web Resources), Markup Languages (HTML, XML, XHTML, XPath) and Document Object Model.

Chapter Three: explains the main details of Website Sitemap, Metadata, and Web Data Extraction as well as Web search engine components are covered.

Chapter Four: describes the proposed Multilateral Web Indexing Model (MWIM) and explains the proposed constructing metadata and sitemap. Then it presents the main data structures types using Search Engine Agent (SEA), which are divided into three modules (Scraper module, Core module, Reporting module) with explaining the algorithms that are building it.

Chapter Five: shows the proposed approach implementation and discusses the results.

Chapter Six: summarize the main conclusions of this thesis and presents an outlook for future works that can improve it.

CHAPTER TWO

WWW CONCEPTS

2.1 Introduction

WWW is an architectural framework for accessing linked documents spread out over millions of machines all over the Internet [1, 9, 10]. Its idea was introduced by Tim Burners-Lee in 1989 at first [11]. Much progress has been made about the Web and related technologies in the past two decades. Web 1.0, Web 2.0, Web 3.0 and Web 4.0 are introduced such as four generations of the Web since the advent of the Web [12].

The first implementation of the Web represents the Web 1.0, which, according to Berners-Lee, could be considered the "read-only Web." In other words, the early Web allowed the users to search for information and read it [11]. The second implementation of the Web represents the Web 2.0 that was defined by Dale Dougherty in 2004 as a read-write Web. One of the outstanding features of Web 2.0 is to support collaboration and to help to gather collective intelligence rather than Web 1.0 [12].

The third implementation is Web3.0 (Semantic Web) where Tim Berners-Lee defines the Semantic Web as follows: "The Semantic Web is an extension of the current Web in which information is given well-defined meaning, better enabling computers and people to work in cooperation" [9]. The Semantic Web is envisioned as the next generation of the Web in which information is machine readable, and automated agents can retrieve, extract, and combine information from the Web [9].

Finally, there is no exact idea about Web 4.0 and its technologies, but it is obvious that the Web is moving toward using artificial intelligence to become as an intelligent Web [12].

2.2 Basics of WWW Technologies

The Web's implementation follows a standard client-server model. In this model, a user relies on a program (called the client) to connect to a remote machine (called the server) where the data is stored. Navigating through the Web is done utilizing a client program called the browser [1].

A **client** is any process that requests specific services from server processes. A **Server** is a process that provides requested services for clients. Client and server processes can reside in the same computer or in different computers connected by network [13].

Web servers and Web browsers are communicating and interacting by client-server computer programs for distributing documents and information over the Internet.

This interaction begins when a user accesses a Website by writing the URL in the Web browser or by clicking a hyperlink in another Web page. Immediately, a request is sent to the Web server using the HyperText Transfer Protocol (HTTP) and the Web server returns the requested object. Figure (2.1) shows the interaction between them [14].

Apache, *Tomcat* and *IIS* are popular Web server programs, and *IE* and *Firefox* are popular Web browsers [10, 14].

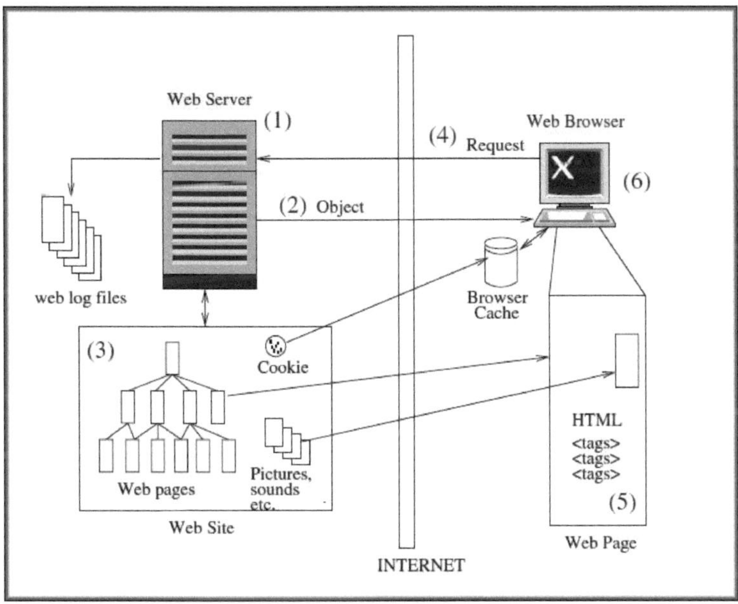

Figure (2.1) The Web Server - Web Browser Interaction

The Web server (1) is a program that runs in the background that receives the request (4) through a specific port, usually the port 80. This program manages a structured set of files, known as a Website (3). Part of these file are Web pages that contain links to other types of data files, such as pictures, sounds, movies or other pages. The main function of the Web server is to fulfill Web pages requests. A Web page (5) is written using the Hyper Text Markup Language (HTML). This consists of a sequence of orders called tags concerning methods on how to display in the user screen the objects (2) requested by the browser (6) and how to retrieve other objects from Web server. These orders are interpreted by the browser which shows the objects on the user screen [14].

2.2.1 Web Browser

Web browser is a software application that enables users to view and interact with information and media files on the Web [15]. A variety of vendors offer commercial browsers that interpret and display a Web document, and all use nearly the same architecture [1].

The basic functions of a Web browser include [16]:

1. Interpreting HTML markup and presenting documents visually;
2. Supporting hyperlinks in HTML documents so the clicking on such a hyperlink can lead to the corresponding HTML file being downloaded from the same or another Web server and presented;
3. Using the HTTP protocol to send requests and data to the Web server and download HTML documents.

2.2.2 Web Server

The term Web server can refer to either the hardware (the computer) or the software (the computer application) that helps to deliver Web content that can be accessed through the Internet. Web content lives on Web servers [16]. The Web server is mainly for receiving document requests and data submission from Web browsers through the HTTP protocol on top of the Internet's TCP/IP layer [16, 17]. The main function of the Web server is to feed HTML files to the Web browsers [17]. Also, the Web server performs the following common tasks as shown in Figure (2.2) [16]:

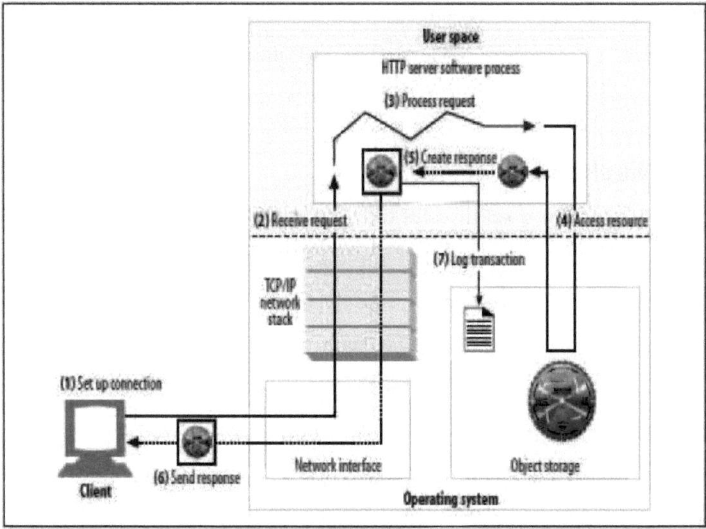

Figure (2.2) Web Server Tasks

1. Set up connection: accept a client connection, or close if the client is unwanted.

2. Receive request: read an HTTP request message from the network.

3. Process request: interpret the request message and take action.

4. **Access resource**: access the resource specified in the message.

5. **Construct response**: create the HTTP response message with the right headers.

6. Send response: send the response back to the client.

7. Log transaction: place notes about the completed transaction in a log file.

When users access a Website, all transactions between the user's browser and the Web server software are logged in ASCII format in server log files. The main log file is usually known as a server transfer log or access log and typically contains fields such as the IP address from which a user's request originated, the time of the request, the method of the request, a numerical code called the status code, indicating the response from the server, and the size in bytes of the transaction [18].

2.2.3 Protocols

A protocol is a set of rules that govern data communications [10]. Web browsers interact with Web servers with a protocol called HTTP (HyperText Transfer Protocol), the hypertext transfer protocol is a request, response protocol with the main purpose of sending files from the server to a client [19]. Which runs on top of TCP/IP network connections. When people click on the submit button of an HTML form or a hyperlink in a Web browser, a TCP/IP virtual communication channel is created from the browser to the Web server specified in the URL; an HTTP GET or POST request is sent through this channel to the destination Web application, which retrieves data submitted by the browser user and composes an HTML file; the HTML file is sent back to the Web browser as an HTTP response through the same TCP/IP channel; and then the TCP/IP channel is shut down[16, 18].

2.2.4 Addressing

A Web server program runs multiple Web applications hosted in different folders under the Web server program's document root folder. A server computer may run multiple server programs, each server program uses a port number (by default a Web server uses port 80) and the server computer has an IP address, like 198.105.44.27, as its unique identifier on the Internet. Domain names, like www.pace.edu, are used as user-friendly identifications of server computers, and they are mapped to IP addresses by a Domain Name Server (DNS) [18].

2.2.4.1 IP Address

IP means Internet Protocol. It is a unique 32-bit binary number assigned to a host and used for all communication with the host. An IP address is, as such, generally shown as 4 octets of numbers from 0-255 represented in decimal form instead of binary form. This address referred to as IPv4 (IP version 4) addresses [20].

The next generation of the Internet Protocol, intended to replace IPv4 on the Internet, was eventually named Internet Protocol Version 6(IPv6). The address size was increased from 32 to 128 bits or 16 octets. In this version, the Internet uses 128-bit addresses that give much greater flexibility in address allocation [10].

2.2.4.2 Uniform Resource Locator (URL)

URL is the most common form of a resource identifier that describes the specific location of a resource on a particular server [22, 23].

URLs are a subset of a more general class of resource identifier called a uniform resource identifier, or URI. URIs are a general concept comprised of two main subsets: the uniform resource locator (URL) and uniform resource name (URNs). URLs identify resources by describing where resources are located, whereas the uniform resource name (URNs) identifies resources by name, regardless of where they currently reside [16, 21].

Most URLs follow a standardized format of three main parts as shown in the Figure (2.3):

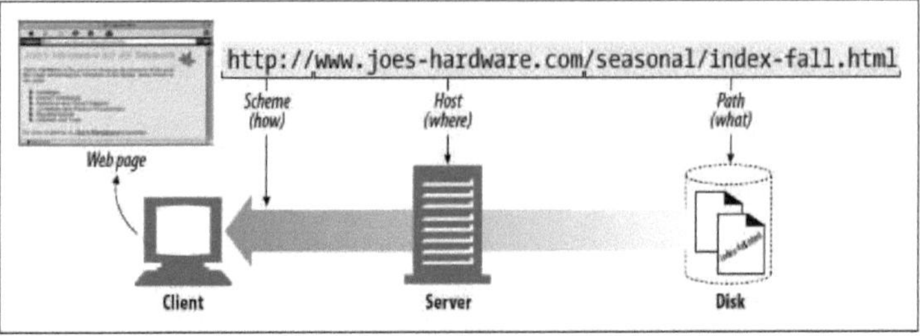

Figure (2.3) A standardized Format of URL

- The first part of the URL (http) is the URL scheme. The scheme tells a Web client how to access the resource. In this case, the URL says to use the HTTP protocol.
- The second part of the URL (www.joes-hardware.com) is the server location. This tells the Web client where the resource is hosted.
- The third part of the URL (/seasonal/index-fall.html) is the resource path. The path tells what particular local resource on the server is being requested [24, 16].

URLs provide a means of locating any resource on the Internet, but these resources can be accessed by different schemes (e.g., HTTP, FTP, SMTP), and URL syntax varies from scheme to scheme. Most URL schemes base their URL syntax on this nine-part general format:

<scheme>://<user>:<password>@<host>:<port>/<path>;<params>?<query>#<frag>

Almost no URLs contain all these components. The three most important parts of a URL are the scheme, the host, and the path. Table (2.1) summarizes the general URL components and Figure (2.4) illustrates an example of these components [16, 23].

Table (2.1) The General URL components	
Component	Description
Scheme /protocol	Which protocol to use when accessing a server to get a resource
user	If the protocol supports the concept of user names, this provides a user name that has access to the resource
password	The password associated with the user name
host	The hostname or IP address of the server hosting the resource
port	The port number on which the server hosting the resource is listening. Many schemes have default port numbers (the default port number for HTTP is 80).
path	The local name for the resource on the server, separated from the previous URL components by a slash (/)
Params/ File	The resource itself
query	Additional information about the resource or the client. Used by some schemes to pass parameters to active applications
frag	A particular location within a resource. The frag field is not passed to the server when referencing the object; it is used internally by the client

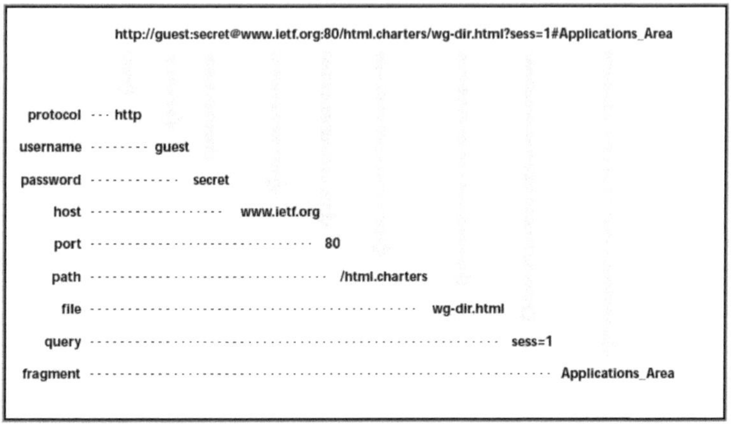

http://guest:secret@www.ietf.org:80/html.charters/wg-dir.html?sess=1#Applications_Area

protocol --- http
username -------- guest
password ------------ secret
host ----------------- www.ietf.org
port ------------------------------ 80
path -------------------------------- /html.charters
file -- wg-dir.html
query -- sess=1
fragment -- Applications_Area

Figure (2.4) URL Component Example

URLs come in two flavors: absolute and relative. An absolute URL is the complete address of a resource and has everything that Web browser needs to find a document and its server on the Web [25].

On the other hand, the relative URL is incomplete, that provides an abbreviated document address, when automatically combined with a "base address" by the Web browser, becomes a complete address (absolute URL) for the document as illustrated in Figure(2.5). Within the relative URL, any component of the URL may be omitted. The browser automatically fills in the missing pieces of the relative URL using corresponding elements of a base URL. This base URL is usually the URL of the document containing the relative URL, but maybe another document specified with the <base> tag. A common form of a relative URL is missing the scheme and server name. Since many related documents are on the same server, it makes sense to omit the scheme and server name from the relative URL [25, 16]. For instance, assume:

The base URL is http://www.joes-hardware.com/tools.html

The relative URL is /hammers.html

By using the base URL, the Web browser can infer the missing information in the relative URL. The relative URL doesn't have the scheme or host. The Web browser can infer that the scheme is http and the host is www.joes-hardware.com.Then:

The absolute URL is: http://www.joes-hardware.com /hammers.html

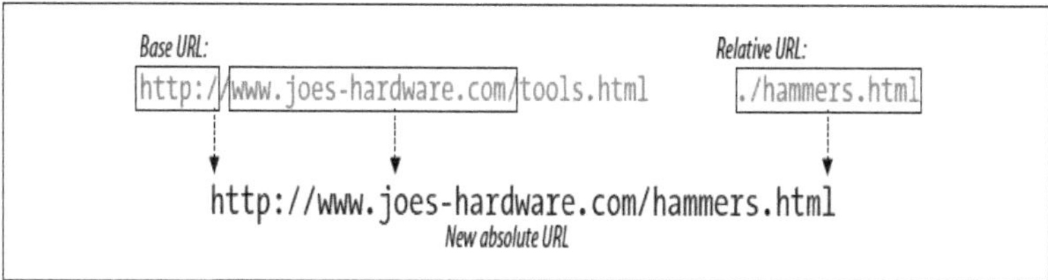

Figure (2.5) Using a Base URL

2.2.4.3 Domain Name System (DNS)

The Domain Name System (DNS) is a hierarchical distributed naming system for computers, services, or any resource connected to the Internet for the mapping of domain names, such as www.google.com, to IP Addresses such as "216.239.51.99" [10, 20, 23].

The DNS is based on a hierarchical and logical tree structure called the domain name space as illustrated in Figure (2.6). The tree has a root at the top. Each node in the tree has a label, which is a string with a maximum of 63 characters. The root label is a null string (empty string). DNS requires that children of a node (nodes that branch from the same node) have different labels, which guarantee the uniqueness of the domain names. Each node in the tree has a domain name. A full domain name is a sequence of labels separated by dots (.)[10].

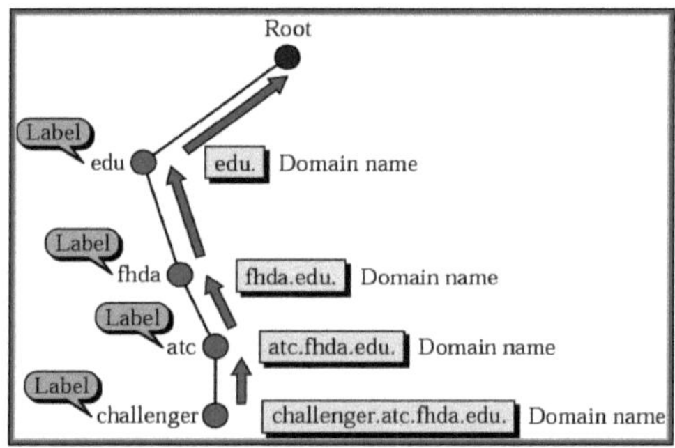

Figure (2.6) The Domain Names and Labels

Domain names ending in .com, .net, or .org are part of the generic top-level domains class. Table (2.2) provides the most popular domain name extensions of generic top-level domains along with their descriptions [26].

Table (2.2) The most popular domain name extensions of generic top-level domains

DOMAIN	Description
.com	Commercial organizations
.net	Network support centers
.org	Nonprofit organizations
.edu	Educational institutions
.gov	Government institutions

2.2.5 Web Resources

Web servers host Web resources. A Web resource is the source of Web content. The simplest kind of Web resource is a static file on the Web server's file system. These files can contain anything: they might be text files, HTML files, Microsoft Word files, Adobe Acrobat files, JPEG image files, or any other format one can think of. However, resources don't have to be static files. Resources can also be software programs that generate content on-demand [16].

A vast proportion of existing Web resources are HTML documents [18]. The documents in the WWW can be grouped into three broad categories: static, dynamic, and active. The category is based on the time at which the contents of the document are determined [10].

a. Static Documents

Static documents are fixed-content documents that are created and stored in a server. The client can get only a copy of the document. In other words, the contents of the file are determined when the file is created, not when it is used as shown in Figure (2.7) [10].

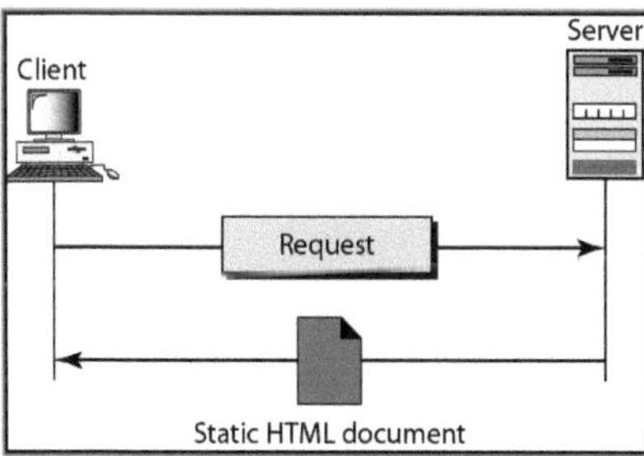

Figure (2.7) Static HTML Document

b. Dynamic Documents

A dynamic document is created by a Web server whenever a browser requests the document. When a request arrives, the Web server runs an application program or a script that creates the dynamic document. The server returns the output of the program or script as a response to the browser that requested the document. Because a fresh document is created for each request, the contents of a dynamic document can vary from one request to another as shown in Figure (2.8) [10].

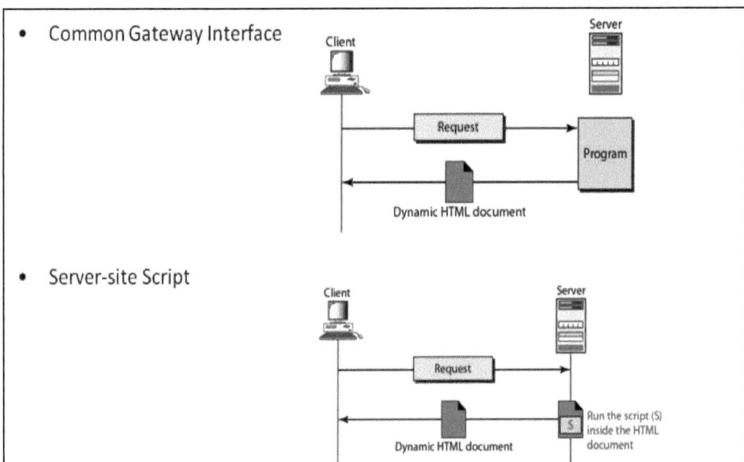

Figure (2.8) Dynamic Documents: CGI and Scripting

c. Active Documents

For many applications, a program or a script is needed to be run at the client site. These are called "active documents". For example, suppose one wants to run a program that creates animated graphics on the screen or a program that interacts with the user. The program definitely needs to be run at the client site where the animation or interaction takes place. When a browser requests an active document, the server sends a copy of the document or a script. The document is then run at the client (browser) site as shown in Figure (2.9) [10].

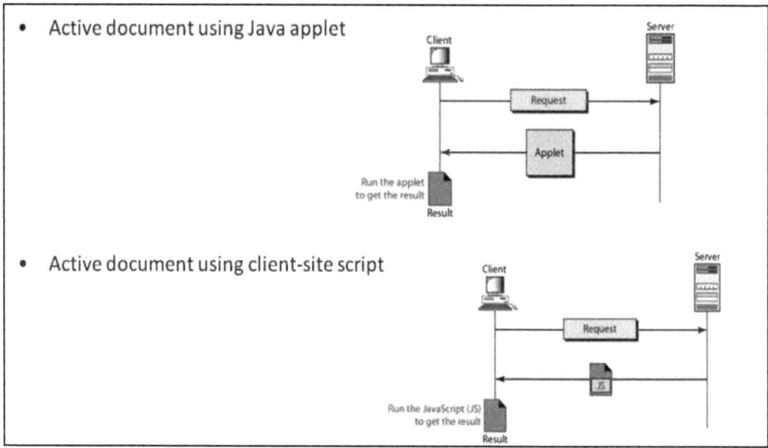

Figure (2.9) Active Documents: Java Applets and Javascript

2.2.6 Search engines

Search engines are often the first place a Web user will go to when searching for information on the WWW. For a business, this means that if the users wish to have a highly visible net presence, they need their Website to be located by users using a search engine. To be able to improve a Web site's chances of appearing near the top of search results, knowledge of how search engines work is required. The first part of the search engine is the crawler or spider. A crawler is a program that traverses Web pages, downloads them for indexing, and follows the hyperlinks that are referenced on the downloaded pages [2].

Once the spiders have completed the task of finding information on Web pages (and it should be noted that is a task that is never actually completed—the constantly changing nature of the Web means that spiders are always crawling), the search engine must store the information in a way that makes it useful. The pages found by the crawlers are passed to the second part of the search engine: the index. The index contains a copy of every Webpage visited by the crawler. It can take some time for pages to be added to an index. The final part of a search engine is the search engine software, which sifts through the pages in the index to find matches to a search and ranks them. The methods used to do this vary with the different search engines [2]. Figure (2.10) shows search engine steps.

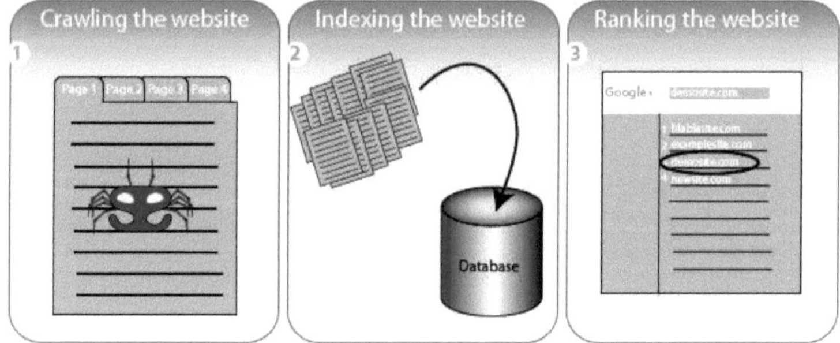

Figure (2.10) Search Engine Steps

2.3 Markup Languages

A coding system represents a set of tags and rules for creating tags that can be embedded in the digital text to provide additional information about the text in order to facilitate automated processing of it, including editing and formatting for display or printing. Markup languages are fundamental to displaying documents in Web browsers [15]. By far the most familiar markup languages to most people are:

2.3.1 HTML

Hypertext Markup Language (HTML) is the language that encodes WWW documents. An HTML is a high level language which mean can be understood by a human, this language allows the full use of hypermedia including text, images, graphics, databases, and sounds, and other types of multimedia [27].

HTML instructions exist within tag. Each Web page consists of n-tags. Tags represent the structure of the Web page and its view to the end-user. An HTML tag name is a predefined keyword, like html, body, head, title, p, and b, all in lower-case. A tag name is used in the form of a start tag or an end tag. A start tag is a tag name enclosed in angle brackets < and >, like <html> and <p>. An end tag is the same as the corresponding start tag except it has a forward slash / immediately before the tag name, like </html> and </p> [28, 29]. A pair of tags and the content are known as an element as shown in Figure (2.11) [30].

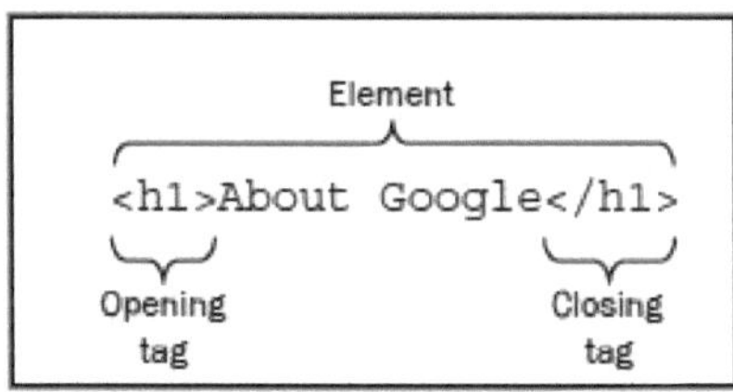

Figure (2.11) HTML Element and Tag

The basic structure of the HTML Web page contains tags of different types. It is structured as illustrated in Figure (2.12).

```
<!DOCTYPE html PUBLIC "-//W3C//DTD HTML 4.01//EN"
    "http://www.w3.org/TR/html4/strict.dtd">
<html>
  <head>
    <title>Page title</title>
  </head>
  <body>
  </body>
</html>
```

Figure (2.12) Basic Structure of HTML Web Page

1. The Doctype

The first item that appears in the source code of a Web page is the doctype declaration. This provides the Web browser (or another user agent) with information about the type of markup language in which the page is written. In the Figure (2.12) above, the doctype relates to HTML 4.01[28].

2. The HTML Element

Immediately after the doctype comes the html element, this is the root element of the document tree and everything that follows is a descendant of that root element [28]. The <html> and </html> tags serve to delimit the beginning and ending of a document. Inside the <html> tag and its end tag are the document's head and body. Since the typical browser can easily infer from the enclosed source that it is an HTML document [25].

3. The HEAD Element

The HEAD element contains information about the document, which is not displayed within the main page itself except the title. This element serves as a container for other elements (such as TITLE, META, BASE elements) [30].

The TITLE Element contains the title of an HTML document, and both the starting and ending tags are required. META tags can be included in HTML document head elements to provide information about the information contained in the page. A common use of META tags is to provide the keywords and descriptions about the page for search engine usage. The BASE element specifies the base URL/target for all relative URLs in a page [25].

4. The BODY Element

The BODY element holds the actual content of the page that is viewed in the browser such as text, formatting, heading, links to other documents,…etc [30].

a) Headings

The need to use the heading tags is to place headings within the body of the document. Usually, the first heading on the page is the same as that in the title. For example: **<H1>Joe Blow's Fabulous Home Page</H1>** will format the words "Joe Blow's Fabulous Home Page" into a large, bold type.

HTML encourages organizing the documents by providing six header tags. The appearance of the header becomes less and less prominent as the number increases [31]:

Tags	Display
<H1> Largest </H1>	**Largest**
<H2>Larger </H2>	**Larger**
<H3>Large </H3>	**Large**
<H4>Small </H4>	**Small**
<H5>Smaller </H5>	**Smaller**
<H6>Smallest</H6>	**Smallest**

b) Text Formatting

A wide range of options exist for the formatting of text, which are important for good design. The most widely used of these tags are <STRONG> (Strong Tag) for logical tags, <B> (Bold Text Tag), and <I> (Italics Text Tag) for physical tags. It should be pointed out that the appearances of text rendered by logical tags are often the same as that of the physical tags. The appearance of text inside these tags will vary depending on the browser interpreting them [30].

c) Links to Other Documents

An HTML file can contain hyperlinks to other Web pages so users can click on them to visit different Web pages. A hyperlink has the general structure of <a href="url">Hyperlink Text</a>. The <A> element is used in HTML to implement links, an essential feature of hypertexts. A link is rather similar to an edge in a directed graph. It connects two objects referred to as anchors. the source anchors are the elements of the document delimited by <A> and </A>. In the WWW, the target anchor is a resource that may be physically stored in the same server as the source document, or maybe in a different server, possibly located in another country or continent. The target resource is identified through a special string called the Uniform Resource Identifier (URI) [18].

2.3.2 XML

XML stands for Extensible Markup Language. A markup language is a language that describes the structure and contents of documents, specifically documents that contain data. The term extensible means capable of being extended and modified. XML is a markup language that can be extended and modified to match the needs of the document author and the data being recorded [32]. XML defines a set of rules for adding markup to data. Markup adds structure to data, and gives a way of talking about the meaning of that data [33]. It has a very strict syntax that makes it a lot easier for computer programs to parse. XML is not an equivalent of HTML though. XML in itself is nothing more than a standard for storing/passing structured data [19].

So, XML and HTML have been designed with different goals [32]:

• XML has been designed to transport and store data, with focus on what data is.

• HTML has been designed to display data, with a focus on how data looks.

2.3.3 XHTML

XHTML combines the flexibility of HTML with the extensibility of XML. XHTML was designed with compatibility in mind. The tags it supports are the same ones, with the same meaning, as those in HTML. Actually, it is even possible for XHTML written keeping a few guidelines in mind to be parsed by HTML parsers. This means XHTML has the strictness and extensibility of XML, while still pertaining to compatibility with the very popular HTML [19].

2.3.4 XPath

XPath (XML Path Language) is a simple query language for navigating around XML trees to identify a set of nodes (for example, elements, attributes, text, and so on) [33]. It's also used in most Document Object Model (DOM) implementations for richer querying capabilities [34]. So, XPath is a very flexible way of working with nodes in a document [35].

2.3.4.1 XPath Expressions

XPath provides a way of describing the location of each node or a set of nodes using a path-like syntax. Technically, these are called "expressions", an expression is the most basic construct of an XPath that returns either a node-set, a string, a Boolean, or a number. Tables (2.3) illustrate the different types of the XPath expressions and operators [35, 36].

Table(2-3) XPath Expressions and Operator

Expression Type	Operator
Location Paths	/ , //, |
Boolean Expressions	or, and
Equality Expressions	=, !=
Relational Expressions	<=, <, >=, >
Expressions	+, -, div, mod, *

One of the most important types of expressions is called the Location Path, which is the syntax used to describe and select nodes in a node document [34]. For example:

- bookstore/book — Selects all book elements that are children of bookstore
- //book — Selects all book elements no matter where they are in the document
- bookstore//book — Selects all book elements that are descendant of the bookstore element, no matter where they are under the bookstore element

Also, the predicates are used to find a specific node or a node that contains a specific value. Predicates are always embedded in square brackets [35]. For example:

• /bookstore/book[last()]	Selects the last book element that is the child of the bookstore element
• /bookstore/book[position()<3]	Selects the first two book elements that are children of the bookstore element
• //title[@lang]	Selects all the title elements that have an attribute named lang
• /bookstore/book[price>35.00]	Selects all the book elements of the bookstore element that have a price element with a value greater than 35.00

2.3.4.2 XPath Functions

XPath also provides a core set of functions that can be used when selecting nodes in a document. There are functions for string values, numeric values, date and time comparison, Boolean values, and more. For example a count ( ) function:

The count ( ) function (Number of nodes) takes a node set, and returns the number of nodes contained in the node set [37].

2.4 Document Object Model

In order to handle a structured document written in HTML or XML, more efficiently and consistently, the WWW Consortium (W3C) has published the Document Object Model (DOM) specification. DOM gives the ability to access and manipulate information stored in a structured HTML or XML document. [38, 39].

DOM is the hierarchal structural representation of data in the form of a tree (nodes), their corresponding parent, brother and child elements, and the node values, often shown pictorially. The hierarchal structure enables the users to identify the tags, their levels, and values in the original document, thus helping to understand the structure of HTML or XML document [35]. Figure (2.13) illustrates a simple HTML document and its corresponding DOM tree. In this example, <BODY> node has three children: element nodes <B> and <I>, and text node #and. Element node <B> has a text node child #bold, and element node <I> has a text node #italic. Following the DOM convention, the <> is to indicate element node, and # to indicate text node [38].

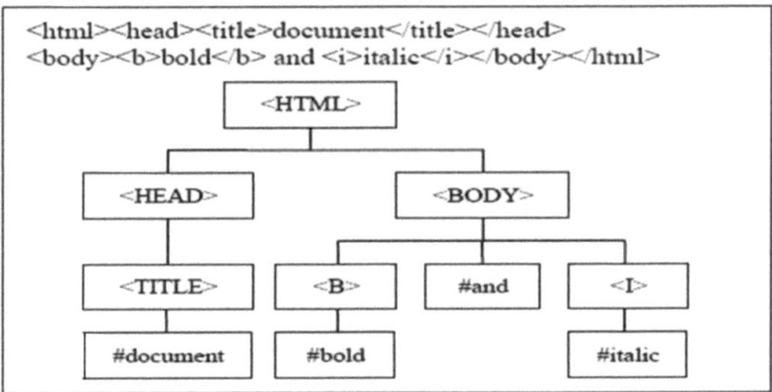

Figure (2.13) A Simple HTML Document and Its DOM Tree

The DOM document structure is organized as a tree, so there are two ways to traverse nodes: depth-first and breadth-first. The depth-first algorithm visits nodes starting at the root node of a subtree and recursively processing each child node of the root node. Based upon where one places the processing logic within the recursive function, the algorithm processes either parent or child nodes first [35]. The standard tree traversing algorithms:

1. Depth-first traversals

In this way, parent nodes are processed first then child nodes are processed Depth-first traversals follow the pattern shown in Figure (2.14). The dotted arrows show the order in which the nodes are visited. In this example, the nodes are visited in the order A, B, D, E, C, F, G [35].

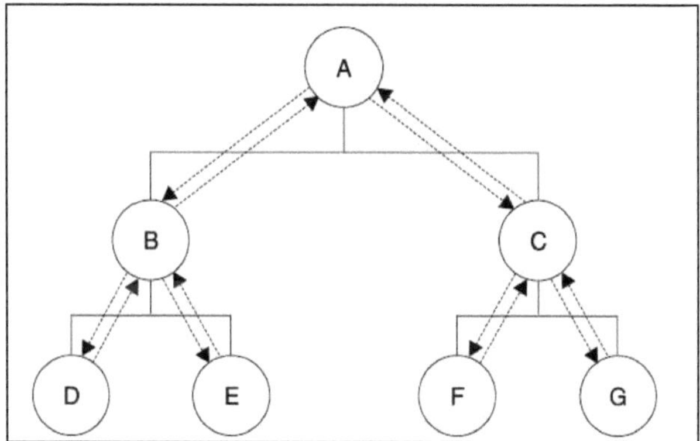

Figure (2.14) Depth-First Traversals

2. Breadth-first traversal

Breadth-first node traversal visits each child node of a given parent before processing any other child nodes of the children of that parent. If that sounds confusing, Figure (2.15) should clear things up. The dotted arrows show the order in which the nodes are visited. In this example, the nodes are visited in the order A, B, C, D, E, F, G [35].

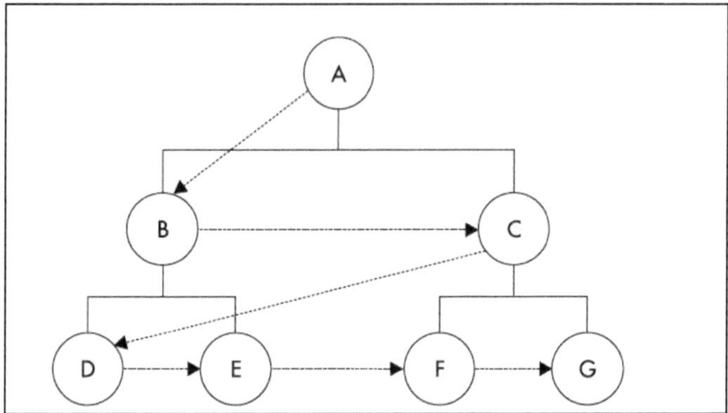

Figure (2.15) Breadth-First Traversal

There is no real logic-related advantage to using breadth-first over depth-first, or vice versa, even though the breadth-first algorithm suffers from a performance disadvantage because it iterates over the child nodes of a given node twice. The application should use the algorithm that is most appropriate to solve one's specific traversal problem [35].

CHAPTER THREE

WEBSITE AND SEARCH ENGINE

3.1 Introduction

The good news about the Internet and its most visible component, the WWW, is that there are hundreds of millions of pages available, waiting to present information on an amazing variety of topics. The bad news about the Internet is that there are hundreds of millions of pages available, most of which titled according to whom of their author, almost all of them sitting on servers with cryptic names. When people need to know about a particular subject, how do we know which pages to read? If somebody is like most people, he/she visits an Internet search engine [2].

This chapter will give an overview of the Website, introduces the Sitemap concepts, describe XML Sitemap, metadata, and Web data extraction.

The remaining expands the search engine concepts and additional details about how the search engine works.

3.2 Website

A Website, indicated by the homepage URL, denotes a set of pages that form a complicated directed graph with ‘‘page” nodes and ‘‘link” edges. In general, a Website can be regarded as a set of URLs sharing the same domain name (i.e. the Website indicates a host that contains pages whose URLs share the same hostname) [40].

3.2.1 Sitemap

A sitemap is a list of URLs of a Website accessible to crawlers or users. So, sitemaps are divided into two broad categories: those created for human users and those specifically created for search engine crawlers. HTML Sitemaps are the organized collection of site links and their associated descriptions, it is designed for the users to help them find content on the page [26].

Over the years, search engines have realized the benefit of Sitemaps. Google has jumped on this concept in 2005 with the creation of its own Google Sitemap Protocol. During 2006, Google Sitemaps Protocol was named XML Sitemap Protocol, to acknowledge its “universal” acceptance. XML Sitemaps are written only for Web spiders. The work of these joint efforts is now under the auspices of Sitemaps.org [26].

3.2.2 XML Sitemap

XML sitemap protocol is an XML file that lists URLs for a site along with additional metadata about each URL (when it was last updated, how often it usually changes, and how important it is, relative to other URLs in the site) so that search engines can more intelligently crawl the site. The premise of using the XML Sitemap protocol was that it would help search engines index content faster while it is providing ways to improve the existing crawling algorithms. Using XML sitemap protocol does not guarantee anything in terms of better page rankings. Furthermore, the use of the XML sitemap protocol is not mandatory for all sites.

It is important to use XML sitemaps because they help Web spiders find site's links. The following points discuss some of the reasons for using sitemaps [26]:

- **Crawling frequency**: One of the biggest benefits of using sitemaps is in timely crawls or re-crawls of site (or just specific pages). XML sitemap documents tell crawlers how often they should read each page.
- **Crawl augmentation**: Use information from sitemaps to augment the crawl and discovery processes especially if the site content is not easily discoverable by the crawler (such as pages accessible only through a form).
- **Poor linking site structure:** Sites with poor linking structures tend to index poorly. Orphan pages, deep links, and search engine traps are culprits of poor site indexing. The use of sitemaps can alleviate these situations, at least temporarily, to give the user enough time to fix the root of the problem.
- **Page priority**: XML sitemap allows assigning a specific priority value for each URL in the XML sitemap file giving search engines suggestions about the importance of each page.
- **Large sites:** Using sitemaps for large sites is important. Sites carrying tens of thousands (or millions) of pages typically suffer in indexing due to deep linking problems. Sites with this many documents use multiple sitemaps to break up the different categories of content.

3.2.3 XML Sitemap Format

The XML sitemap protocol format consists of XML tags. The file itself must be UTF-8 encoded. The Sitemap must begin with an opening <urlset> tag and end with a closing </urlset> tag.

Each link in an XML Sitemap can have <url> tag as a parent XML tag and four attributes as children entry for each <url> parent tag. The first attribute, loc, is the URL location and it is mandatory. The rest of the attributes are optional and include lastmod, changefreq, and priority [26]. Figure (3.1) is an example of a Sitemap file with a single link and the available XML tags are described in the table (3.1).

```
<?xml version='1.0' encoding='UTF-8'?>

<urlset xmlns="http://www.sitemaps.org/schemas/sitemap/0.9">
  <url>
      <loc>http://mydomain.com/</loc>
     <lastmod>2010-01-01</lastmod>
     <changefreq>weekly</changefreq>
      <priority>0.5</priority>
  </url>
</urlset>
```

Figure (3.1) XML Sitemap Example

Table (3.1) XML Sitemap Tags

no	Attribute	Description	
1	<urlset>	Mandatory	Encapsulates the file and references the current protocol standard
2	<url>	Mandatory	Parent tag for each URL entry
3	<loc>	Optional	Represents the actual URL or link

			value
4	< lastmod>	Optional	Represents the time and date when a particular link was last modified
5	< changefreq>	Optional	Represents a hint as to how often a particular link might change with a range of values, including: always, hourly, weekly, monthly, yearly, never.
6	<priority>	Optional	Represent the priority of the URL relative to other URLs on site. Valid values range from 0.0 to 1.0.

3.3 Metadata

The term metadata, purportedly first used in 1969 is often called 'data about data' [41, 15] or 'information about information'. The term 'meta' is derived from the Greek word denoting a nature of a 'higher order' or more 'fundamental kind', or 'above', 'beyond', and 'of something in a different context' [42].

A metadata record consists of a number of pre-defined elements representing specific attributes of a resource, and each element can have one or more values. These elements could be the content, quality, condition, and other characteristics of data or other pieces of information which presents the best way to share information about the data without having to provide the actual data. In other words, metadata is structured information that describes, explains, locates, or otherwise makes it easier to retrieve, use, or manage an information resource also states that metadata is the key to ensuring that (digital) resources will survive and continue to be accessible into the future [41].

3.3.1 Web-related Metadata

Metadata standards for describing Internet resources have appeared (i.e. representation of Web-related metadata), such as meta tags, the Dublin Core Metadata Element Set (DCMES), and the Resource Description Framework (RDF) [51]. Some types of Web metadata are explained in the following:

1. Meta Tags

Some search engines originally popularized the use of two simple metadata elements, "keywords" and "description," that can be easily and invisibly embedded in the <HEAD> section of Web pages by their authors using the HTML meta tag [15, 17]. Here is an example:

<META NAME="KEYWORDS" CONTENT="data standards, metadata, Web resources, WWW, cultural heritage information, digital resources, Dublin Core, RDF, Semantic Web">

<META NAME="DESCRIPTION" CONTENT="Version 3.0 of the site devoted to metadata: what it is, its types and uses, and how it can improve access to Web resources; includes a crosswalk.">

The original intention was that the "keyword" metadata could be used to provide more effective retrieval and relevance ranking, whereas the "description" tag would be used in the display of search results to provide an accurate, authoritative summary of the particular Web resource [15].

2. Dublin Core

The Dublin Core Metadata Initiative (DCMI) created its first metadata standard, the Dublin Core Metadata Element Set (DCMES), to facilitate search and retrieval of Web-based resources [41]. The DCMES is a set of fifteen information elements that can be used to describe a wide variety of resources for simple cross-disciplinary resource discovery. The fifteen elements are Contributor, Coverage, Creator, Date, Description, Format, Identifier, Language, Publisher, Relation, Rights, Source, Subject, Title, and Type [15, 41].

3. Resource Description Framework

The RDF is a standard developed by the WWW Consortium (W3C) for encoding resource descriptions (i.e. metadata) in a way that computers can "understand," share, and process in useful ways [43]. A resource can be a Web page, an entire Website, or any item on the Web that contains information in some form. RDF metadata is normally encoded using XML [43, 44]. That enables the exchange, and reuse of structured metadata [45]. RDF is a graph framework to represent the information and to give metadata about the resources on the WWW [46, 47, 48]. The RDF model consists of three major components [45, 47]:

a. **Resources**: All things being described by RDF expressions are called resources. A resource may be an entire Web page, part of a Web page, an entire Website, or an object that is not directly accessible via the Web page.

b. **Properties:** A property is a specific aspect, characteristic, attribute, or relation used to describe a resource. Each property has a specific meaning, defines its permitted values, the types of resources it can describe, and its relationship with other properties.

c. Statements: A specific resource together with a property plus the value of that property for that resource is an RDF statement. These three individual parts of a statement are called the subject, predicate, and object, of the statement, respectively. More concrete details can be found in the following example.

For the statement: "The owner of the Website http://www.asu.edu is Arizona State University".
Figure (3.2) shows RDF to express this statement using (1) a resource or subject (http://www.asu.edu), (2) a property name or predicate (owner), and (3) an atomic value or object (Arizona State University).

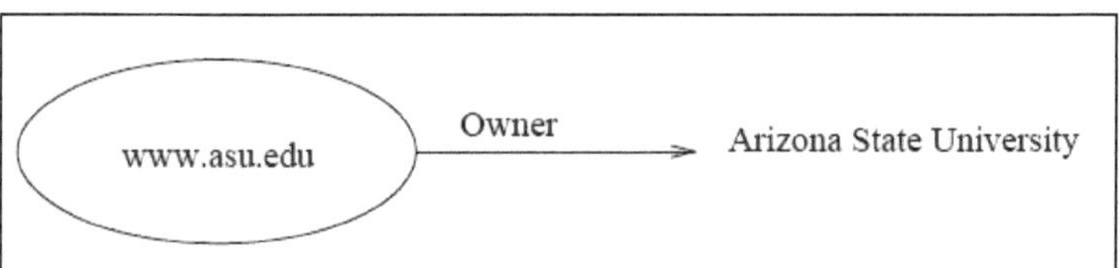

Figure (3.2) RDF of a Statement Example

3.3.2 Metadata Associate Model

There are various ways to associate metadata with resources [49] as illustrated in Figure (3.3), and explained in the following:

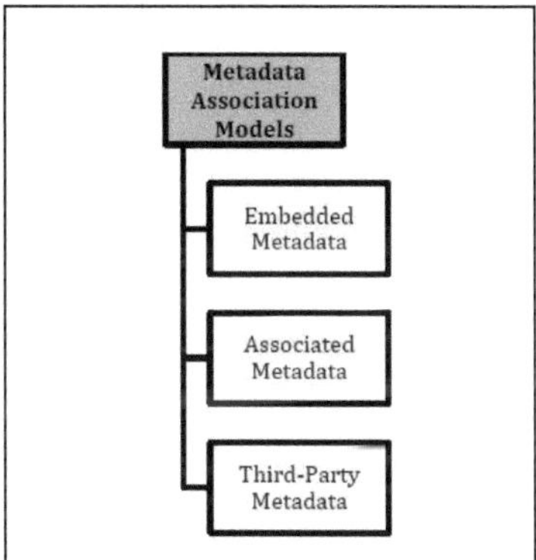

Figure (3.3) Metadata Associate Model

1. **Embedded metadata** resides within the markup of the resource. This implies that the metadata is created at the time that the resource is created, often by the author. Experts differ concerning whether author-created metadata is best or whether it is better to have trained practitioners evaluate and describe resources. As a practical matter, resource description expertise is a scarce and costly commodity, and thus any investment by authors in the description of their intellectual products is likely to be of value [49].

2. **Associated metadata** is maintained in files tightly coupled to the resources they describe. The advantage of associated metadata derives from the relative ease of managing the metadata without altering the content of the resource itself, but this benefit is purchased at the cost of simplicity, necessitating the co-management of resource files and metadata files [49].

3. **Third-Party metadata** is maintained in a separate repository by an organization that may or may not have direct control over or access to the content of the resource. Typically such metadata is

maintained in a database that is not accessible to harvesters, though the emerging Open Archives Initiative Metadata Harvesting Protocol proposes a system that encourages the disclosure of metadata repositories [49].

3.3.3 Attributes and Characteristics of Metadata

In addition to different metadata types and associate model, metadata exhibits many different characteristics. Some key characteristics of metadata are presented in the following [15]:

1. Source of metadata

a) Internal metadata generated by the creating agent for an information object at the time when it is first created or digitized.

b) External metadata relating to an original item or information object, which is created later, often by someone other than the original creator.

2. Method of metadata creation

a) Automatic metadata generated by a computer.

b) Manual metadata created by humans.

3. Nature of metadata

a) Non-expert metadata created by persons who are neither subject specialists nor information professionals, e.g., Meta tags created for a personal Web page.

b) Expert metadata created by subject specialists and/or information professionals, often not the original creator of the information object.

4. Status

a) Static metadata that does not or should not change once it has been created.

b) Dynamic metadata that may change with use, manipulation, or preservation of an information object.

5. Structure

a) Structured metadata that conforms to a predictable standardized or proprietary structure.

b) Unstructured metadata that does not conform to a predictable structure.

6. **Semantic**

a) Controlled metadata that conforms to a standardized vocabulary or authority form, and that follows standard content (i.e., cataloging) rules.

b) Uncontrolled metadata that does not conform to any standardized vocabulary or authority form.

3.4 Web Data extraction

Web data extraction systems are a broad class of software applications targeting at extracting information from Web sources like Web pages [50]. A Web data extraction system usually interacts with a Web source and extracts data stored in it: for instance, if the source is a HTML Web page, the extracted information could consist of elements in the page as well as the full-text of the page itself. Eventually, extracted data might be post-processed, converted in the most convenient structured format, and stored for further usage [50].

Unlike Information Retrieval (IR), which is concerned with how to identify relevant documents from a document collection, Web data extraction produces structured data ready for post-processing, which is crucial to many applications of Web mining and searching tools [51].

Web data extraction is closely related to Web indexing, which indexes information on the Web using a Web crawler and is a universal technique adopted by most search engines [52].

The importance of Web data extraction systems depends on the fact that, today, a large (and quickly growing) amount of information is continuously produced, shared, and consumed online: Web Data extraction systems allow to efficiently collecting this information with a limited human effort [50].

One of the most exploited features in Web data extraction is the semi-structured nature of Web pages. These can be naturally represented as labeled ordered rooted trees, where labels represent the tags proper of the HTML mark-up language syntax, and the tree hierarchy represents the different levels of nesting of elements constituting the Web page. The representation of a Web page by using a labeled ordered rooted tree is usually referred to as DOM. The general idea behind the DOM is that HTML Web pages are represented using plain text, which contains HTML tags, so as free-text. HTML tags may be nested one into another, forming a hierarchical structure. This hierarchy is captured in the DOM by the document tree, whose nodes represent HTML tags.

The document tree (DOM tree) has been successfully exploited for Web Data extraction purposes in number of techniques, such as addressing elements in the document tree: XPath. One of the main advantages of the adoption of the Document Object Model (DOM) for the HTML language is the possibility of exploiting some tools typical of XML languages. In particular, the XML Path Language (XPath) provides a powerful syntax to address specific elements of a DOM Tree in a simple manner [50].

3.5 Web Search Engines

Web search engines are very important tools for people to get the desired information on the Web [53, 54]. Web search engine is a program that collects data taken from the content of files available on the Web and puts them in an index or database that Web users can search in a variety of ways [15]. The search results provide links back to the pages matching the user's search in their original location. The currency of the results is purely based on when the search engine's crawlers visited the site and brought back the meta information to the search engine's database. A crawler must crawl the Web as fast as possible to serve the user more efficiently [53].

The typical design of search engines is a "cascade", in which a Web crawler creates a collection that is indexed and searched. This is a cascade model, in which operations are executed in strict order: first crawling, then indexing, and then searching [55, 56]. A basic model of a typical search engine is given in figure (3.4).

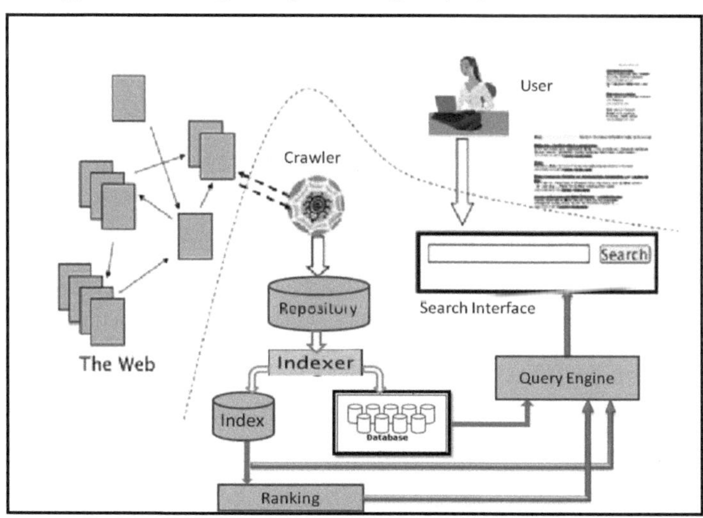

Figure (3.4) Structure of Web Search Engine

The architecture of a search engine includes a crawler, indexer, query engine, and search interface. The crawler is a software program that traverses the Web by following links and downloading Web pages that are sent to the indexer, which creates the search index. The search index is organized as an inverted-file data structure, which can be likened to a book's index. Information about hyperlinks is stored in a link database, giving quick access to a Web page's out links through its URL. The query engine processes the query by first retrieving information about relevant Web pages from the search and then combining this information to provide a ranked list of result pages. The search interface is responsible for displaying the results on the user's browser [57].

A summary of a Web-page as created by the search engine can be seen as meta-data. If every site had reliable meta-data in a standardized format, the work of a search engine would be made a lot easier [19].

3.5.1 Web Crawler

The WWW can be seen as a directed graph. The vertices represent the Web pages and the directed edges represent the links from one page to another. A Web crawler is a program that uses this graph structure of the Web (the Web-graph) to visit Web-pages in an automated manner by following the edges from every page in the Web-graph. This process is called crawling the Web. In their infancy, these programs were also called wanderers, robots, spiders, walker, and worm [35]. Web spiders crawl not only Web pages, but also many other files, including robots.txt, sitemap.xml, and so forth [26]. The general processes that a crawler takes are as follows [58,59]:

- Check for the next page to download – the system keeps track of pages to download in a "queue".
- Links are checked before being added to the frontier.
  - URL duplicates.
  - Content duplicates and nearly-duplicates.
  - robots policy checking (robots.txt file)

  Check to see if the page is "allowed" to be downloaded - checking a “robots exclusion” file and also reading the header of the page to see if any exclusion instructions are provided to do this. Some people don't want their pages to be archived by search engines.
- Links normalization.
- Download the whole page.
- During parsing, the text (with HTML tags) is passed on to the indexer, so are the links contained in the page (links analysis).
- Save the summary of the page and update the "last processed" date for the page so that the system knows when it should re-check the page at a later date.

A crawler begins by taking a URL from the frontier and fetching the Web page at that URL, generally using the HTTP protocol. The fetched page is then written into a temporary store, where the number of operations are performed on it. Next, the page is parsed and the text as well as the links in it is extracted. The text (with any tag information – e.g., terms in boldface) is passed on to the indexer. Link information including anchor text is also passed on to the indexer for use in ranking [6, 59, 60].

In addition, each extracted link goes through a series of tests to determine whether the link should be added to the URL frontier. First, the crawler tests whether a Web page with the same content has already been seen at another URL. The simplest implementation for this would use a simple fingerprint such as a checksum [6, 59].

Next, a URL filter is used to determine whether the extracted URL should be excluded from the frontier based on one of several tests. For instance, the crawl may seek to exclude certain domains (say, all .com URLs) – in this case the test would simply filter out the URL if it were from the .com domain [6, 59].

Many hosts on the Web place certain portions of their Websites off-limits to crawling, under a standard known as the robots exclusion protocol. Robots exclusion this is done by placing a protocol file with the name robots.txt at the root of the URL hierarchy at the site. Here is an example robots.txt file that specifies that no robot should visit any URL whose position in the file hierarchy starts with /yoursite/temp/ [6, 59].

User-agent: *

Disallow: /yoursite/temp/

The star character (*) implies all Web spiders. The robots.txt file must be fetched from a Website to test whether the URL under consideration passes the robot restrictions, and can, therefore, be added to the URL frontier. Rather than fetching it afresh for testing on each URL to be added to the frontier, a cache can be used to obtain a recently fetched copy of the file for the host. This is especially important since many of the links extracted from a page fall within the host from which the page is fetched and therefore can be tested against the host's robots.txt file. Thus, by performing the filtering during the link extraction process, there should be especially high locality in the stream of hosts that is needed to test for robots.txt files, leading to high cache hit rates. Unfortunately, this runs afoul of Webmaster's politeness expectations. A URL (particularly one referring to a low-quality or rarely changing document) may be in the frontier for days or even weeks. It was essential to perform the robots filtering before adding such a URL to the frontier, its robots.txt file could have been changed by the time the URL is dequeued from the frontier and fetched. Robots-filtering must be frequently performed before attempting to fetch a Web page [59].

As it turns out, maintaining a cache of robots.txt files is still highly effective; there is sufficient locality even in the stream of URLs dequeued from the URL frontier [6, 59].

URL normalization next, a URL should be normalized in the following sense: often the HTML encoding of a link from a Web page p indicates the target of that link relative to the page p. Thus, there is a relative link encoded thus in the HTML of the page en.wikipedia.org/wiki/Main_Page:

<ahref="/wiki/Wikipedia:General_disclaimer"
title="Wikipedia:General
disclaimer">Disclaimers</a>

points to the URL:

http://en.wikipedia.org/wiki/Wikipedia:General_disclaimer.

Finally, the URL is checked for duplicate elimination: if the URL is already in the frontier or already crawled, it is not added to the frontier. When the URL is added to the frontier, it is assigned a priority based on which it is eventually removed from the frontier for fetching [6, 59].

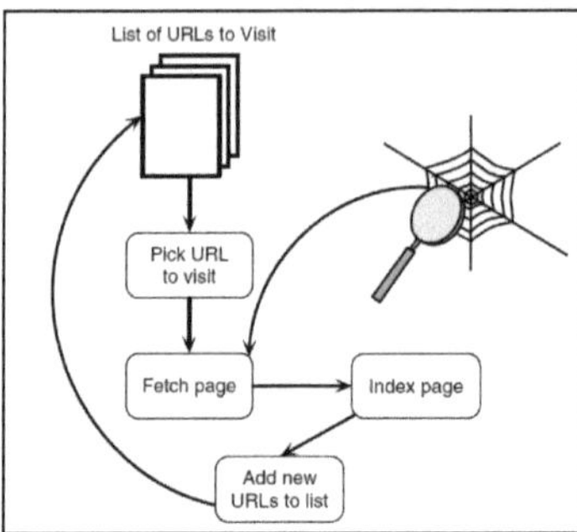

Figure (3.5) Basic Crawler Architecture

Each Web search engine follows its unique timetable for re-crawling the Web and updating its content collection [61]. Figure (3.5) illustrate the basic crawler architecture.

3.5.2 Indexer

Indexing of Web content is a challenging task assuming an average of 1000 words per Web page and billions of such pages [62]. Indexing is a process of organizing a search engine's database of Web pages (collects, parses, and stores data) to facilitate fast and accurate information retrieval [21, 60]. An indexer is a program that "reads" the pages, which are downloaded by spiders; each search engine uses a variety of indexing algorithms to sort the contents of a page. In general, the occurrence of keywords, and the contents of html elements like meta tags, titles, and headers and looks for words that are considered important words in bold, italics or header tags are given more priority, are all taken into account as the index is created retrieval efficiency [61, 62].

Indexing starts with parsing the Website content using a parser. Any parser, which is designed to run on the entire Web, must handle a huge array of possible errors. The parser can extract the relevant information from a Web page by excluding certain common words (such as a, an, the - also known as stop words), HTML tags, Java Scripting, and other bad characters. A good parser can also eliminate commonly occurring content in the Website pages such as navigation links, so that they are not counted as a part of the page's content. Once the indexing is completed, the results are stored in memory, in a sorted order. This helps in retrieving the information quickly. Indexes are updated periodically as new content is crawled. Some indexes help create a dictionary (lexicon) of all words that are available for searching. Also, a lexicon helps in correcting mistyped words by showing the corrected versions in a search result. A part of the success of the search engine lies in how the indexes are built and used. Various algorithms are used to optimize these indexes so that relevant results are found easily without much computing resource usage [62]. The main activities of search engine indexer are:

1. Document Conversion

It is the first step in the translating process which refers to any method that transfers an input text into a sequence of character codes. Documents are converted, because there are many different formats with different definitions of how text is represented. Document conversion can then be characterized as a translation of text from an arbitrary format into a target sequence of codes that can easily be manipulated [6].

2. Parsing or Tokenization

Tokenization is the extraction of plain words and terms from a document, stripping out administrative metadata and structural or formatting elements, e.g. removing HTML tags from the HTML source file for a Web page. This operation needs to be performed before indexing or before converting documents to vector representations that are used for retrieval or categorization [18].

The case of HTML documents is relatively easy. The simplest approach consists of reducing the document to an unstructured representation, i.e. a plain sequence of words with no particular relationship among them other than serial order. This can be achieved by only retaining the text enclosed between <html> and </html>, removing tags, and perhaps converting strings that encode international characters to a standard representation [18].

An HTML parser must tolerate errors and include recovery mechanisms to be of any practical usefulness, obviously, after the plain text is extracted, punctuation and other special characters need to be stripped off. In addition, the character case may be folded (e.g. to all lowercase characters) to reduce the number of index terms [18].

3. Stop Word

Stop word remover removes the high frequent terms that do not depict the context of any document. These words are considered unnecessary and irrelevant for the main content to Web pages. Words like a‘, ‘an‘, ‘the‘, ‘of‘‘, ‘and‘ etc. that occur in almost every text are some of the examples for stop words[63].

Reducing index space and improving performance are important reasons for eliminating stop words [24].

4. Stemming

Stemming is a technique for removing the morphological component from the term, thus reducing the word to the base form. It is normally achieved by using a rule-based approach, usually based on suffix stripping [57, 63, 64].

This does not only means that different variants of a term can be conflated to a single representative form – but it also reduces the dictionary size in the index, that is, the number of distinct terms needed for representing a set of documents. A smaller dictionary size results in a saving of storage space and processing time [64].

The effect of stemming is an increase in the number of documents retrieved for a given keyword due to the increase in the posting list for that keyword, and thus the recall (the number of relevant documents retrieved divided by the total number of relevant documents) of the system is also increased. This means that the results list returned by the query engine will be much larger, but it does not necessarily imply that the results list will contain a higher proportion of relevant pages, that is, the precision of the system may decrease [57].

A partial stemmer is often used, which focuses only on plurals and the most common suffixes such as ED and ING, but the standard stemming algorithm for English language is the Porter Stemmer algorithm [63]. The definition of the algorithm, as it appears in Porter stemming paper [65] is:

A \consonant\ in a word is a letter other than A, E, I, O, or U, and other than Y preceded by a consonant. (The fact that the term `consonant' is defined to some extent in terms of itself does not make it ambiguous.) So in TOY, the consonants are T and Y, and in

SYZYGY they are S, Z and G. If a letter is not a consonant it is a \vowel\.

A consonant will be denoted by c, a vowel by v. A list ccc... of length greater than 0 will be denoted by C, and a list vvv... of length greater than 0 will be denoted by V. Any word, or part of a word, therefore has one of the four forms:

CVCV ... C, CVCV ... V, VCVC ... C, and VCVC ... V

These may all be represented by the single form

[C]VCVC ... [V]

Where the square brackets denote arbitrary presence of their contents. Using (VC) {m} is to denote VC repeated m times, this may again be written as [C] (VC) {m} [V]. m will be called the \measure\ of any word or word part when represented in this form. The case m = 0 covers the null word. Here are some examples:

m=0 TR, EE, TREE, Y, BY.

m=1 TROUBLE, OATS, TREES, IVY.

m=2 TROUBLES, PRIVATE, OATEN, ORRERY.

The \rules\ for removing a suffix will be given in the form

(Condition) S1 ⟶ S2

This means that if a word ends with the suffix S1, and the stem before S1 satisfies the given condition, S1 is replaced by S2. The condition is usually given in terms of m, e.g.

(m > 1) EMENT ⟶

Here S1 is `EMENT' and S2 is null. This would map REPLACEMENT to REPLAC, since REPLAC is a word part for which m = 2.The ‘condition’ part may also contain the following:

*S - the stem ends with S (and similarly for the other letters).

v - the stem contains a vowel.

*d - the stem ends with a double consonant (e.g. -TT, -SS).

*o - the stem ends cvc, where the second c is not W, X or Y (e.g. -WIL, -HOP). And the condition part may also contain expressions with \and\, \or\ and\not\, so that c (m>1 and (*S or *T)) tests for a stem with m>1 ending in S or T, while (*d and not (*L or *S or *Z)) tests for a stem ending with a double consonant other than L, S or Z. Elaborate conditions like this are required only rarely. In a set of rules written beneath each other, only one is obeyed, and this will be thc one with the longest matching S1 for the given word. For example, with

SSES ⟶ SS, IES⟶ I, SS⟶SS, S ⟶

(Here the conditions are all null) CARESSES maps to CARESS since SSES is the longest match for S1. Equally CARESS maps to CARESS (S1=`SS') and CARES to CARE (S1=`S').

In the rules below, examples of their application, successful or otherwise, are given on the right in lower case. The algorithm now follows:

Step 1a

SSES ⟶ SS	caresses ⟶ caress
IES ⟶ I	ponies ⟶ poni
	Ties ⟶ ti
SS ⟶ SS	caress ⟶ caress
S ⟶	cats ⟶ cat

Step 1b

(m>0) EED ⟶ EE	feed ⟶ feed
	Agreed ⟶ agree
(*v*) ED ⟶	plastered ⟶ plaster
	Bled ⟶ bled
(*v*) ING ⟶	motoring ⟶ motor

sing ⟶ sing

If the second or third of the rules in Step 1b is successful, the following is done:

AT ⟶ ATE conflat(ed) ⟶ conflate

BL ⟶ BLE troubl(ed) ⟶ trouble

IZ ⟶ IZE siz(ed) ⟶ size

(*d and not (*L or *S or *Z)) ⟶ single letter

hopp(ing) ⟶ hop

tann(ed) ⟶ tan

fall(ing) ⟶ fall

fizz(ed) ⟶ fizz

(m=1 and *o) ⟶ E fail(ing) ⟶ fail

fil(ing) ⟶ file

The rule to map to a single letter causes the removal of one of the double letter pair. The -E is put back on -AT, -BL and -IZ, so that the suffixes -ATE, -BLE and -IZE can be recognized later. This E may be removed in step 4.

Step 1c

(*v*) Y ⟶ I happy ⟶ happi

Sky ⟶ sky

Step 1 deals with plurals and past participles. The subsequent steps are much more straightforward.

Step 2

(m>0) ATIONAL ⟶ ATE relational ⟶ relate

(m>0) TIONAL ⟶ TION conditional ⟶ condition

(m>0) ENCI ⟶ ENCE valenci ⟶ valence

(m>0) ANCI ⟶ ANCE hesitanci ⟶ hesitance

(m>0) IZER ⟶ IZE digitizer ⟶ digitize

(m>0) ABLI ⟶ ABLE conformabli ⟶ conformable

(m>0) ALLI → AL	radicalli → radical
(m>0) ENTLI → ENT	differentli → different
(m>0) ELI → E	vilely → vile
(m>0) OUSLI → OUS	analogously → analogous
(m>0) IZATION → IZE	vietnamization → vietnamize
(m>0) ATION → ATE	predication → predicate
(m>0) ATOR → ATE	operator → operate
(m>0) ALISM → AL	feudalism → feudal
(m>0) IVENESS → IVE	decisiveness → decisive
(m>0) FULNESS → FUL	hopefulness → hopeful
(m>0) OUSNESS → OUS	callousness → callous
(m>0) ALITI → AL	formaliti → formal
(m>0) IVITI → IVE	sensitiviti → sensitive
(m>0) BILITI → BLE	sensibiliti → sensible

The test for the string S1 can be made fast by doing a program switch on the penultimate letter of the word being tested. This gives a fairly even breakdown of the possible values of the string S1. It will be seen in fact that the S1-strings in step 2 are presented here in the alphabetical order of their penultimate letter. Similar techniques may be applied in the other steps.

Step 3

(m>0) ICATE → IC	triplicate → triplic
(m>0) ATIVE →	formative → form
(m>0) ALIZE → AL	formalize → formal
(m>0) ICITI → IC	electriciti → electric
(m>0) ICAL → IC	electrical → electric
(m>0) FUL →	hopeful → hope
(m>0) NESS →	goodness → good

Step 4

(m>1) AL ⟶ revival ⟶ reviv

(m>1) ANCE ⟶ allowance ⟶ allow

(m>1) ENCE ⟶ inference ⟶ infer

(m>1) ER ⟶ airliner ⟶ airlin

(m>1) IC ⟶ gyroscopic ⟶ gyroscop

(m>1) ABLE ⟶ adjustable ⟶ adjust

(m>1) IBLE ⟶ defensible ⟶ defens

(m>1) ANT ⟶ irritant ⟶ irrit

(m>1) EMENT ⟶ replacement ⟶ replac

(m>1) MENT ⟶ adjustment ⟶ adjust

(m>1) ENT ⟶ dependent ⟶ depend

(m>1 and (*S or *T)) ION ⟶ adoption ⟶ adopt

(m>1) OU ⟶ homologou ⟶ homolog

(m>1) ISM ⟶ communism ⟶ commun

(m>1) ATE ⟶ activate ⟶ activ

(m>1) ITI ⟶ angulariti ⟶ angular

(m>1) OUS ⟶ homologous ⟶ homolog

(m>1) IVE ⟶ effective ⟶ effect

(m>1) IZE ⟶ bowdlerize ⟶ bowdler

The suffixes are now removed. All that remains is a little tidying up.

Step 5a

(m>1) E ⟶ probate ⟶ probat

rate ⟶ rate

(m=1 and not *o) E ⟶ cease ⟶ ceas

Step 5b

(m > 1 and *d and *L) ⟶ single letter

controll ⟶ control

roll ⟶ roll

The algorithm is careful not to remove a suffix when the stem is too short, the length of the stem being given by its measure, m. There is no linguistic basis for this approach. It has been merely observed that m could be used quite effectively to help decide whether or not it is wise to take off a suffix. For example, in the following two lists:

list A	list B
RELATE	DERIVATE
PROBATE	ACTIVATE
CONFLATE	DEMONSTRATE
PIRATE	NECESSITATE

-ATE is removed from the list B words, but not from the list A words. This means that the pairs DERIVATE/DERIVE, ACTIVATE/ACTIVE, DEMONSTRATE/DEMONSTRABLE, ECESSITATE/NECESSITOUS, will conflate together. The fact that no attempt is made to identify prefixes can make the results look rather inconsistent. Thus PRELATE does not lose the -ATE, but ARCHPRELATE becomes ARCHPREL. In practice, this does not matter too much, because the presence of the prefix decreases the probability of an erroneous conflation. Complex suffixes are removed bit by bit in the different steps. Thus GENERALIZATIONS is stripped to GENERALIZATION (Step 1), then to GENERALIZE (Step 2), then to GENERAL (Step 3), and then to GENER (Step 4). OSCILLATORS is stripped to OSCILLATOR (Step 1), then to OSCILLATE (Step 2), then to OSCILL (Step 4), and then to OSCIL Step 5). The algorithm is given in Figure (3.6) [64].

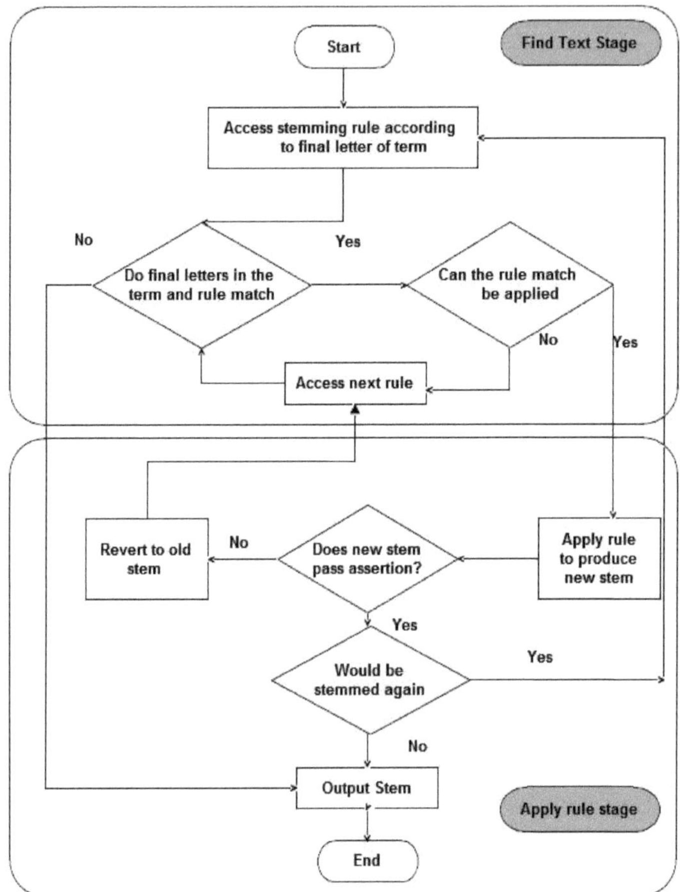

Figure (3.6) Porter Stemmer Flowchart

5. Search Index

The search index is a data repository containing all the information the search engine needs to match and retrieve Web pages. The type of data structure used to organize the index is known as an inverted file as shown in Figure (3.7). It is very much like an index at the back of a book [57]. It contains all the words appearing in the Web pages crawled, listed in alphabetical order (this is called the index file), and for each word, it has a list of references to the Web pages in which the word appears (this is called the posting list) [57, 61].

Often, more information is stored for each entry in the index such as the number of documents in the posting list for the entry, that is, the number of Web pages that contain the keyword, and for each individual entry in the posting file we may also store the number of occurrences of the keyword in the Web page and the position of each occurrence within the page. This type of information is useful for determining content relevance [57].

The search index will also store information pertaining to hyperlinks in a separate link database, which allows the search engine to perform hyperlink analysis, which is used as part of the ranking process of Web pages. The link database can also be organized as an inverted file in such a way that its index file is populated by URLs and the posting list for each URL entry, called the source URL, contains all the destination URLs forming links between these source and destination URLs. The link database for the Web can be used to reconstruct the structure of the Web and to have good coverage; its index file will have to contain billions of entries. When we include the posting lists in the calculation of the size of the link database, then the total number of entries in the database will be an order of magnitude higher. Compression of the link database is thus an important issue for search engines, who need to perform efficient hyperlink analysis [57].

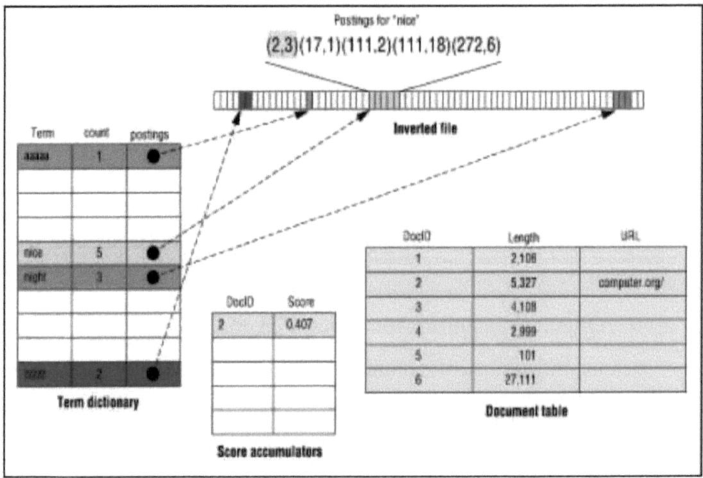

Figure (3.7) Inverted index file and associated data structures

3.5.3 Query Engine

The query engine is the algorithmic heart of the search engine. The query engine receives the search requests from users then processes it in two steps. In the first step, it takes the query submitted by the user, splits the query into terms, and retrieve from the search index information about potentially relevant Web pages that match the terms in the user query and the second step, a ranking of the results is produced, from the most relevant downwards by calculating a similarity score between the query and each document. The higher the similarity score, the higher the ranking for a particular document. The ranking algorithm combines content relevance of Web pages and other relevant measures of Web pages based on link analysis and popularity. Different ranking algorithms can produce very different document rankings. These ranking algorithms are extremely important to both search engines and users [57, 61], for example:

The search engine Google uses an algorithm called the Page Rank algorithm. The Page Rank algorithm gives each page an importance rank. This rank is based on how many pages link to those pages. It doesn't only look at the number of links to a page though, but also at the rank of the "voting" pages themselves. This means that a page gets a high score if a lot of high score pages link to the page. This system functions that good that Google is at the present the number one search engine [19].

This ranking which carries valuable, discriminatory power is arrived at by combining two scores, the content score (derived using on-site parameters) and the popularity score (derived using off-site factors)[66].

3.5.4 Search Interface

Once the query is processed, the query engine sends the results list to the search interface, which displays the results on the user's screen. The user interface provides the look and feel of the search engine, allowing the user to submit queries, browse the results list, and click on chosen Web pages for further browsing. From the usability point of view, the users can distinguish between sponsored links, which are ads, and organic results, which are ranked by the query engine. While most of the Web search engines have decided to move away from the Web portal look toward the simpler, cleaner look pioneered by Google, several of them, notably Yahoo and MSN, maintain the portal look of their home page, offering their users a variety of services in the hope of converting them to customers, independently of their use of the core search services [57].

CHAPTER FOUR

MULTILATERAL WEB INDEXING MODEL

(MWIM)

4.1 Introduction

Due to the large extent of information and the diversity of information sources available these days on the WWW, a user is more likely to the help of a search engine to find the desired information. Indexes are important information-finding tools that can enhance Website appearance in the search bar. The purpose of storing an index is to optimize speed and performance in finding relevant documents for a search query. Perhaps most importantly, site indexes provide direct access to chunks of information without the need for traversing multiple links in a hierarchy.

The indexing process requires a lot of computational power, and also a significant amount of I/O, this means most of the work is done before any searches are performed. Because most of the hard work has already been done while indexing, searches are super-fast.

A new metadata structure is proposed in this thesis to create a different Web indexing model by using traditional methods (DOM with XPath). This chapter will detail the main components of the Multilateral Web Indexing Model (MWIM) with an explanation of the methods and algorithms used to construct each one.

4.2 Proposed MWIM

This thesis proposes a different approach for a search engine to retrieve the relevant data from a Website by establishing a tied collaboration between the search engine and the Websites served by it. Since almost all modern Websites are dynamic Websites overlaying on a server side scripting, these are capable of auto-generating metadata describing Website contents. The proposed solution requires for a Website to expose two kinds of metadata: **Sitemap** and **Content Extract** file. This is done by a server-side component called **Search Engine Agent** which acts as a bridge between a Website and the search engines. Figure (4.1) illustrates an overview of the data path in the proposed approach and all the involved processes in a MWIM. The **Search Engine Agent** is responsible for generated metadata, this metadata contains structured information of a Website and contents of each Web page.

This approach has been called **Multilateral Web Indexing Model**, abbreviated as **MWIM** due to the joint involvement of both: search engines and Websites in Web indexing activities. **MWIM** is characterized by:

- Distributed, where the indexing is distributed on many Website servers instead of search engine's one;
- Multilateral, where the Website and the search engine are collaborated to perform the indexing process;
- Full Text Index;
- Containing XML Sitemap with the same properties and new elements;
- Standard Stemmer for English language and stop word list including 1300 word.

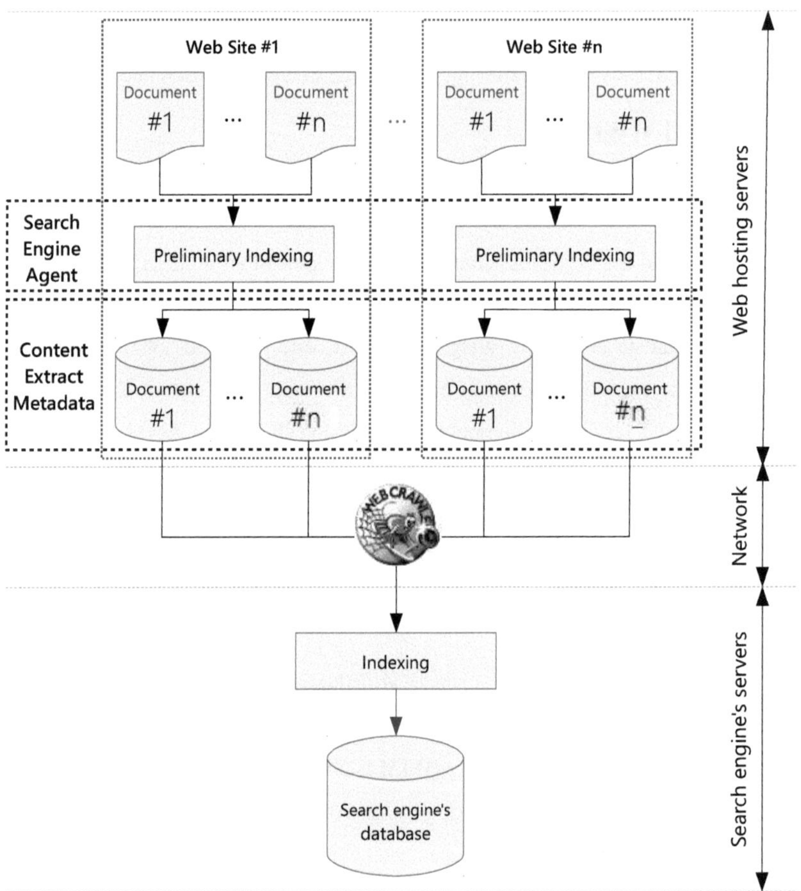

Figure (4.1) An overview of the data path and the involved processes in a MWIM

4.3 MWIM Components

The Multilateral Web Indexing Model construction relies upon two components: the Search Engine Agent and the metadata exposed to the search engines.

As shown in Figure (4.2), the Search Engine Agent working as event-driven software, waiting for the Website's back end calling to start metadata generating process. The Website's back end is a program already found in any Website for managing its contents. For example the Website-back end of a Content Management System may call the Search Engine Agent after the Web page is modified or created, to generate its metadata.

The auto-generated metadata of a Web page contents has been called **Content Extract (CE)** and this metadata would contain : the textual representation extracted from the actual markup format of the Web page, semantic information retrieved from the mark-up and additional metadata. CE's are serialized in individual binary files with (.ce) extension, along of the other Website's files, being able to be publicly retrieved via a URL address. These files keep almost the same information like the original document, but in a way independent of the original file format. So, the search engine after retrieves the CE, can know what was in the source document, even if the search engine does not support to scrape that file format.

The CE files are stored and not sent directly to the search engines to provide a continuous availability of the metadata, instead of re-computing it each time it is requested. A simple representation of the path in which the data of Websites is following along with the incurred processes in MWIM is explained in Figure (4.1).

Along of the Content extract metadata, the **sitemap** is another member of the exposed metadata. The purpose of a sitemap is to describe the structure of a

Website by specifying the location of each Web page along with other information such as the modification date and the CE location. Thus, the proposed sitemap enforce their presence as being the only one source in determining the structure of a Website. Thus the content addition, modification, or removal on WWW would be reflected very shortly in the search engine's indexes because the search engines are aware of them. Unlike the present use of sitemaps, which are not mandatory and only used as a helper for the Web crawlers.

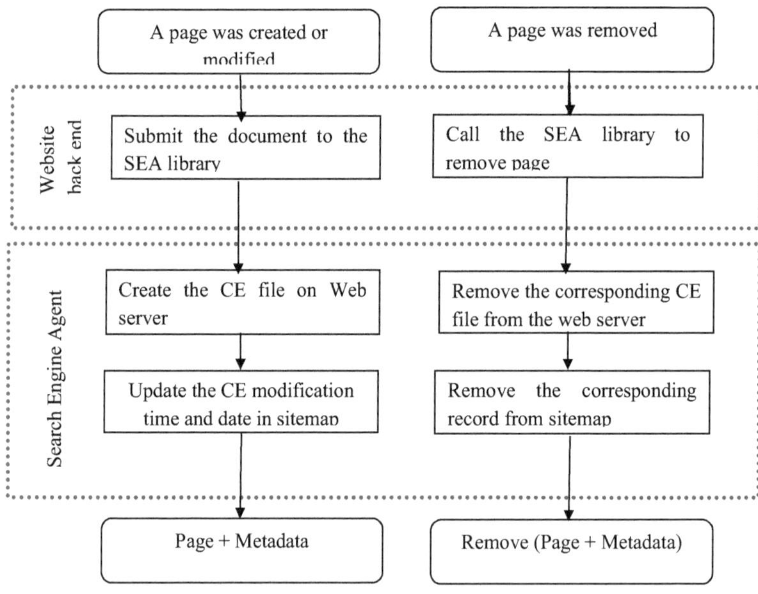

Figure (4.2) A diagram of the processes executed by the back end of a Website and SEA for the creation, modification and removal of a Web resource.

4.4 The Generated Metadata Types

MWIM approach is designed to generate different metadata files for each Website where the CE file will be generated for each Web page, while a single sitemap file will be generate for the whole Website. According to section (3.3.3) in the previous chapter, the proposed metadata characterized by:

1. **Source of metadata/** Internal metadata.

2. **Method of metadata creation/** Automatic metadata.

3. **Nature of metadata/** Expert metadata.

4. **Status/** Dynamic metadata.

5. **Structure/** Structured metadata.

6. **Semantic/** Controlled metadata.

The details and components of each metadata file types are explained in the following subsections:

4.4.1 Content Extract Metadata

Basically, the proposed Content Extract binary file can be considered as a forward index, i.e. contain data that can be directly indexes for all search engines without any other processing. The benefits from using binary format are:

1. Smaller size - to reduce the bandwidth consumption.
2. Faster loading - allows faster indexing of pages by the search engine.

The structure of the whole information included in this file can be classified as three main parts: (Document properties, Textual content and Non-textual content), as presented in Figure (4.3) and explained in the following subsections.

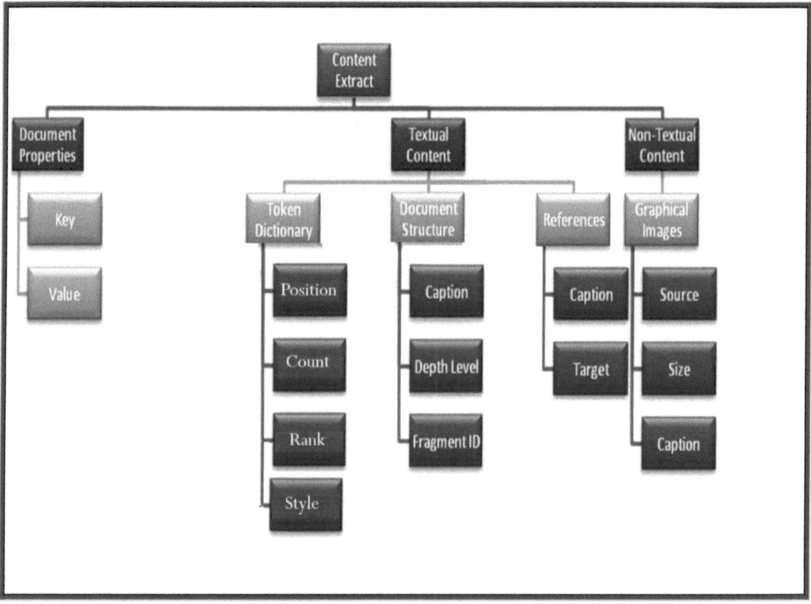

Figure (4.3) An overview of the information contained in CE

4.4.1.1 Document Properties

Document properties consist of the information that describes the Web page document to which the generated Content Extract belongs, such as the title, the author name, the description, the keywords, etc. This information is organized as a **property list** called **Document Properties**. This property list is represented as a pair of particular information called **property value**, and a unique name called **property key** which is used to identify each property. The basic set of properties are enumerated in table (4.1):

Table (4.1) Document property set	
Key	Description
`title`	The title of the document.
`url`	The URL address of the document.
`base-url`	The base URL address used to resolve relative URL references, if it is different than the URL of the document, or if the URL of the document is not available.
`author`	The name of the document's author.
`description`	A short description of the document.
`keywords`	A list of words which describes the topic of the document.
`retrieval-date`	The date and time when the Content Extract was retrieved, in the ISO-8601 representation and converted to the UTC offset.
`language`	The IETF language tag of the natural language in which the content of the document is written. If the content of a document is written in multiple natural languages or dialects, a separate Content Extract file should be created for each particular language or dialect.

Since this information is stored in a property list, without a predefined structure, additional property schemes may be defined, added and used without requiring any change to the Content Extract's file format.

4.4.1.2 Textual Content

This part contains the information that describes the textual content of the document. This is divided into three categories:

- **Token Dictionary** – a tokenized representation of the document text along with additional information such as formatting styles.
- **Document Structure**– the index of a document as defined by its headings.
- **References**–a list of references (links) to the same or other document(s).

1. The Token Dictionary

The Token Dictionary consists of a **token** list extracted from the document text, where each token has associated information which describes its context in the document. A token is a stemmed representation of a natural language word.

The information associated to a token consists of a/an:

- **Position**– an enumeration of the offsets where a token is located in the text. This offset represents the ordinal number of the token in text.
- **Occurrence Count**– represents the count of the ocurrences of a token in the text.
- **Rank**— the rank of the token determined by its context in the document. The ranking criteria are presented in Table (4.2). The rank may only be raised, relative to the rank of the portion of the text which a token is subordinated to. Thus, the tokens from a portion of empathised text from a hyperlink will still be ranked to 5, even if the rank of the empathised text is 3. The rank of this

token will keep the same level like the portion of text which is subordinated to (the hyperlink) because a rank downgrade is not possible.

- **Formatting Style**– describes the basic formatting style appears in the tokens of the document, such as: font face, size, style (bold, italic and underlined) and color.

The tokens of a Token Dictionary are pre-filtered to eliminate the stop-words, to improve the compactness of the Content Extract file.

Table 4.2 – A list of criteria for raising the rank of a token

Rank	Criteria for raising the rank of a token
10	Part of the document's title.
9	Represents one of the document's keywords.
8	Part of the document's description.
6	Part of a reference (hyperlink).
5	Part of a subtitle.
4	Part of the caption of a graphical image.
3	Part of bold text.
2	Part of a paragraph.

2. Document Structure

Basically the document structure represents a table of document contents infered from subtitles (or headings). Each element of this table is described through its:

- **Caption**– the caption of a subtitle, in the original natural language form.
- **Depth Level**– a subtitle may be subordinated to another in a hierarchical tree. This represents the hierarchical level of the element relative to the tree's root, and relative to the previous elements as relationship. As an example: elements `A`, `B`, and `C` have the respective depth levels 1, 2 and 1. Their relationship is determined by the previous elements, a greater depth level than the previous element will indicate a descendent relationship, while a lesser one, an ascendant relationship respectively a sibling relationship if the depth level remains the same. Thus, `A` and `C` are siblings, while `B` is descendant of `A`.
- **Fragment ID**– the identifier used on the fragment segment of a URL to refer to the current subtitle, if it is available. As an example, in HTML this identifier coincides with the value of the ID attribute applied to the heading tags.

3. References

Contain a list of references to other documents or fragments of the same document. These references are equivalent to the HTML hyperlinks. Each element of this list consists of:

- **Caption**– the caption of a reference, in the original natural language form.

- **Target**– the destination, absolute or relative, URL address of this reference.

4.4.1.3 Non-Textual Contents

Contain information which describes non-textual elements. Currently, only the **Graphical Image List** is considered. The Graphical Image List describes graphical elements such as Figures or pictures. These elements are described by the following information.

- **Source**– the URL address of the resource, if available.
- **Size**– the size in pixels of the window which displays the graphical image (viewport size), if available.
- **Caption**— the caption text of the graphical image, in the original natural language, if available.

4.4.2 Sitemap of Website

In a real-life implementation, using the XML Sitemap is not mandatory for all Websites but it is made mandatory in proposed approach to promote faster crawling and indexing of the sites. Doing a bit of math along with making some assumptions in example shown below can help to see the bigger picture to the importance of a Sitemap.

<u>Example</u>[26]:

Suppose there are three Websites, each with a different number of pages:

Site A: 100 pages, 1 second average per-page crawler retrieval time.

Site B: 10,000 pages, 1.2 seconds average per-page crawler retrieval time.

Site C: 1 million pages, 1.5 seconds average per-page crawler retrieval time.

Now suppose you have changed one of the pages on all three sites. Here is the worst-case scenario (modified page crawled last) of retrieving this page by a Web crawler:

Site A: 100 pages × 1 second = 100 seconds (1.67 minutes)

Site B: 10,000 pages × 1.2 seconds = 12,000 seconds (3.3 hours)

Site C: 1 million pages × 1.5 seconds = 1.5 million seconds (17.4 days)

When you consider that Web crawlers typically do not crawl entire sites all in one shot, it could take a lot longer before the modified page is crawled. This example clearly shows that big sites are prime candidates for using XML Sitemaps.

4.4.2.1 The Proposed XML Sitemap

The proposed Sitemap represents an extended version of the original XML Sitemaps 0.9 protocol introduced by Google. This extension provides support for exposing information regarding to the Content Extract corresponding to a Sitemap entry such as: the location, last update time and the hash code of the source document. The extension is achieved by allowing new elements belonging to the XML namespace "urn:mwimc/content-extract-manifest" to be descendants of the "url" element defined in the original protocol. The new elements added to the proposed XML Sitemap are:

1. "loc" element:

The location of the Content Extract file is required for the search engine to be able to retrieve the corresponding Content Extract file of a Sitemap entry. It should be specified as an absolute URL through the "loc" element. If a Web page uses multiple natural languages or dialects, a Content Extract should be created for each particular language or dialect. When more than one Content Extract corresponds to a single Web page, each "loc" element should contain the "lang" attribute which specifies the IETF language tag corresponding to the respective Content Extract.

2. "lastupd" element:

The search engine identifies the updated Content Extract files by comparing the actual modification date and time of each Sitemap Entry with the corresponding values from the cached version of the Sitemap. The date and time in ISO-8601 UTC form of the last update procedure for a Sitemap entry are specified through the "lastupd" element.

3. "fingerprint" element

For a client side tool which automatically generates the metadata for static Websites, the MD5 (a 128 bit hash function designed by Ron Rivest in1991, it processes a variable-length message into a fixed-length output of 128 bits [67]) hash code of the source files can be specified to allow the identification of the Web pages which have been modified from the previous update procedure. The search engine compares the modification date of the content extract.

But the client-side tools do not have any reliable way to check for the modification date of the source files, and uses this hash. The hash code should be specified through the "fingerprint" element.

Figure (4.4) shows an exemplification of a XML Sitemap extended with Content Extract Manifest:

```
<?xml version="1.0" encoding="utf-8"?>
<urlset
xmlns="http://www.sitemaps.org/schemas/sitemap/0.9"
xmlns:ce="urn:mwimc/content-extract-manifest">
<url>
<loc>
http://www.samplesite.com/index
</loc>
<cem:loc>
    http://www.samplesite.com/index.ce
</cem:loc>
<cem:lastupd>
    2013-7-29T16:18:12
</cem:lastupd>
<cem:fingerprint>
    721F46B7E21B49D47CCEFAF6BBF4729D320A3AF5
</cem:fingerprint>
</url>
</urlset>
```

Figure (4.4) Example of Proposed Sitemap

4.5 Proto-MWIM SDK

The MWIM Software Development Toolkit (SDK) is a set of development tools and libraries to implement the Proto- MWIM that includes:

- A Search Engine Agent capable to scrape HTML pages and serialize/deserialize a Content Extract with the same specifications presented previously.
- A graphical user interface tool set, called **Proto-MWIM Toolkit**, which represents a graphical shell for allowing a user to manipulate the functionality provided by the Search Engine Agent, directly without the need of any programmatic implementation. These tools will be detailed in chapter 5.

4.5.1 Search Engine Agent

The Search Engine Agent (SEA) prototype consists of a set of component libraries built on .NET Framework 4. The routine which interoperates with these libraries should reside in the back-end of a Website built on ASP.NET technology, or a client side application and it could be called the "caller" routine.

SEA features a modular architecture, to facilitate eventual extensions of its functionality, currently containing three modules as shown in Figure (4.5), and explained below:

- **The HTML Scraper Module**– automatically generates Content Extract metadata files from HTML markup code.

- **The Core Module**– provides an API interface for creating, accessing, serializing and deserializing Content Extract files.
- **The Reporting Module**– generates a report regarding to the information stored in a Content Extract file.

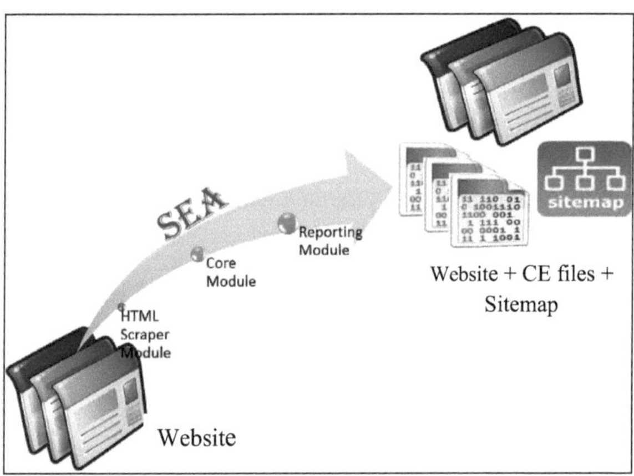

Figure (4.5) Search Engine Agent Modules

4.5.1.1 The HTML Scraper Module

As shown in Figure (4.6), this module retrieves Content Extract by scraping a HTML markup code. The HTML Scraper Module is represented by "ProtoMWIM.Scrapers.HTML.dll".

The scraping procedure represents the parsing of the document. This is achieved by HTML Agility Pack, an open-source HTML parser (.Net library) that scans the HTML page and provides a DOM Tree which then can be used to fetch required values from the page's content.

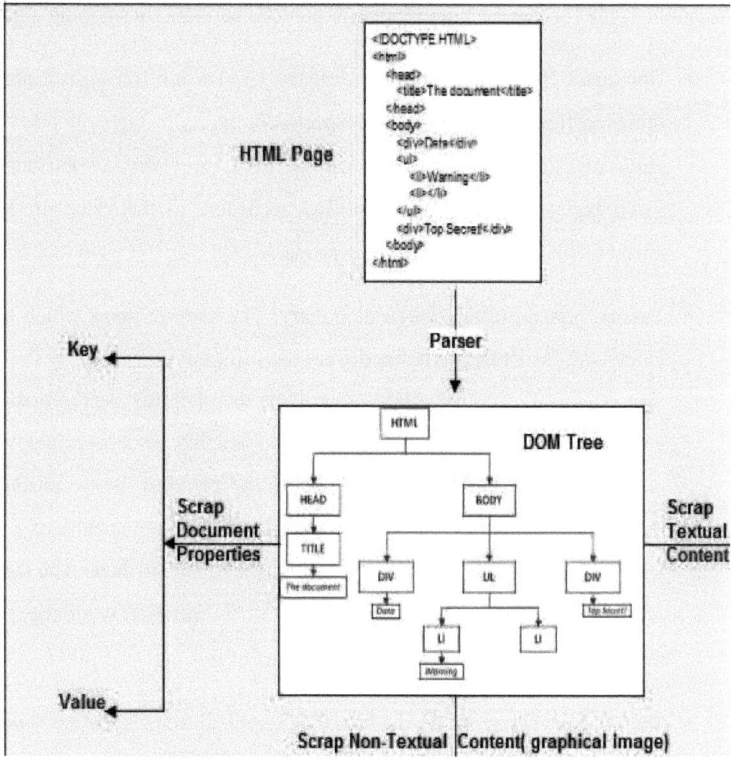

Figure (4.6) HTML Scraper Module

There are two stages to generate Content Extract file:

- **Step1**

To retrieve the value of Document Properties, Textual content except token dictionary and Non-Textual content, the HTML tags which contain the corresponding information are located by performing XPath queries against of the DOM tree. The Xpath queries corresponding to these Document Properties can be found in Table (4.3) and a brief representation can be found in scraping algorithm.

- Document Properties: The value of the `title` and `base-url` property is retrieved from the `title` and respectively the `base` tag contents, while the value of other properties are retrieved from the `content` attribute of the `meta` tag which has been identified according to the value of the `name` attribute which coincides with the property key.
- Textual content except token dictionary: The subtitle items which build the Document Structure are defined from the information carried by the heading tags. The caption of the subtitle represents the plain text representation of its contents. The depth level coincides to the heading level associated with the respective tag, while the ID coincides to the value of the *`id`* attribute. The source of information required to build the References section of a Content Extract represents the `a` tags that contains a `href` attribute. The target of a reference coincides with the value of the *`href`* attribute while the caption is the plain text representation of the contents of this tag.
- Non-Textual Content: As the last, the Graphical Images section is populated with the information carried by the *`img`* tags, which are located using the following query. All required information is available as attributes of this tag, the source corresponding to the *`src`* attribute, the size to the *`width`* and *`height`* attributes, and caption to the *`alt`* attribute.

The value of these properties can be retrieved only if the entire markup code of the HTML document is available. In the case when the scraping is performed by submitting only a fragment of mark-up code without the html tag as the root node, the values of these properties should be filled by the Search Engine Agent caller.

Algorithm 4.1: Scraping Algorithm	
Input Parameter	DOM Tree
Output Parameters	Document Properties, Textual content except token dictionary and Non-Textual content

```
Begin
   Titlement ← /html/head/title
   Metaelement ← /html/head/meta[@name != '' and @content != '']
   Linkelement ← "//a[@href !="
   headelement ←  "//h1|//h2|//h3|//h4|//h5|//h5|//h6"
   imgelement ←  "//img[@src != '']"
       for each Node(element)in DOM Tree DO
         if element= Titlement or Metaelement or Linkelement or
            headelement or imgelement then
            CE file ← element(content)
        end if
      end for
End
```

<table>
<tr><td colspan="4">Table (4.3) The XPath Queries Used for Retrieving the CE</td></tr>
<tr><th></th><th>Property key</th><th>XPath expression for the property value</th><th>Collected information</th></tr>
<tr><td rowspan="4">Document Properties</td><td>Title</td><td><code>/html/head/title</code></td><td>Page Title</td></tr>
<tr><td>Base URL</td><td><code>/html/head/base[@href != '']</code></td><td>Page Base URL</td></tr>
<tr><td rowspan="2">Metadata</td><td rowspan="2"><code>/html/head/meta[@name != '' and @content != '']</code></td><td>Name Attribute</td></tr>
<tr><td>Content Attribute</td></tr>
<tr><td rowspan="5">Textual Content</td><td rowspan="3">Heading</td><td rowspan="3"><code>//h1|//h2|//h3|//h4|//h5|//h6</code></td><td>Caption</td></tr>
<tr><td>Depth level</td></tr>
<tr><td>Fragment ID</td></tr>
<tr><td rowspan="2">Link</td><td rowspan="2"><code>//a[@href != '']</code></td><td>Caption</td></tr>
<tr><td>Target</td></tr>
<tr><td rowspan="3">Non-Textual Content</td><td rowspan="3">Image</td><td rowspan="3"><code>//img[@src != '']</code></td><td>Source</td></tr>
<tr><td>Size</td></tr>
<tr><td>Caption</td></tr>
</table>

- **Step 2**

To retrieve Text, formatting style information and rank them, and send it to next step that represent the tokenization of the text contained by the text node. These step and tokenization processes are responsible of creating the Token Dictionary.

This step is accomplished by traversing the nodes of the Content Elements, and acting according to the type of each node:

- **Formatting style tags**–the current formatting style, which is inherited from the parent node, is altered accordingly. The tags which define the formatting Styles are listed in Table (4.4).
- **Tags involved in ranking criteria**— if the new rank is greater than the one inherited from the parent node, the rank will be raised to the new value; otherwise, it will keep the inherited value. The list of tags that belongs to this category is presented within the Table (4.5) along with their corresponding ranks.
- **Text node**— the value of this node is submitted to the tokenizer along with the formatting style information and the rank inherited from the parent node.

Table (4.4) The Information Source in Retrieving the Information Style

Formatting Style Property	**Source**	
	HTML Tag	**Comment**
Font Face	`font`	Retrieved from the `face` attribute.
Color	`body`	Retrieved from the `text` attribute.

<table>
<tr><td></td><td>font</td><td>Retrieved from the color attribute</td></tr>
<tr><td rowspan="4">Size</td><td>font</td><td rowspan="2">Retrieved from the size attribute</td></tr>
<tr><td>basefont</td></tr>
<tr><td>small</td><td>Decreased by the presence</td></tr>
<tr><td>big</td><td>Increased by the presence</td></tr>
<tr><td rowspan="2">Bold</td><td>strong</td><td rowspan="5">Set by the presence of the tag</td></tr>
<tr><td>b</td></tr>
<tr><td rowspan="2">Italic</td><td>em</td></tr>
<tr><td>i</td></tr>
<tr><td>Underlined</td><td>u</td></tr>
</table>

Table (4.5) The rank level and their corresponding HTML tags

Rank	Corresponding Tags
6	A
5	h1, h2, h3, h4, h5, h6
3	b, strong, i, em
2	P

A depth-first tree traversal algorithm is used for traversal DOM tree, because this traversal order preserves the original flow of the text in Tokenizer, which is required to record the correct token positions and to join correctly the tokens spanned on multiple nodes. The traversal process is explained in HTML DOM Traversal algorithm that is given below:

Algorithm 4.2: HTML DOM Traversal Algorithm	
Input Parameters	
Element	The HTML node instance to process. Its descendant elements will be also processed.
Style	The formatting style of the parent of the specified node. Initially, the value is empty.
rank	The rank level of the parent node. Initially, the value is empty.
Output Parameters	
Text, Rank, Style	Collect the Text from all elements with its rank and style

```
Begin
    newStyle←style
    ! check the kind of the current node
        if isTextElement(element) then
           tokenize(getElementText(element), style, rank)

       ! Collect rank value:
       else if
           getElementName(element) = "h1", "h2", "h3", "h4", ↴
            ↳"h5", "h6" then  newRank ← 5
       else if getElementName(element) = "a" then newRank ← 6
```

```
        else if getElementName(element) = "b", "stroing", "i","em"
        then newRank ← 3
        else if getElementName(element) = "p" then newRank ← 2

      ! Collect style information:
      else if getElementName(element) = "b", "strong" then
          newStyle.bold = true
      else if getElementName(element) = "i", "em" then
          newStyle.italic = true
      else if getElementName(element) = "u" then
          newStyle. Underlined = true
      else if getElementName(element) = "font", "body", "facefont",
         "small", "big" then retreive color and size
  end if
  ! Call recursively the same procedure for the children elements
  ! of the current element.
  for allchild in element
     recall HTML DOM Traversal(child, newStyle, max(rank, newRank))
  end for
End
```

4.5.1.2 The Core Module

The Core Module represents the base component library of the Search Engine Agent which exposes the following functionality to the caller:

- Provides access to the Content Extract information, either by the deserialization of a previously generated file or by creating a blank Content Extract instance.
- Provides text tokenization services, including stop-word removal, stemming, and aggregation of the results in the token dictionary as illustrated in Figure (4.5).

- Performs a process to convert the Content Extract information to format can store or transmit during the network, i.e. serialization.

The publicly exposed API members of this module are representing by the "ProtoMWIM.Core.dll". The basic function in Core Module functions is Text Tokenization.

The Text Tokenization represents the most important functions of a search engine indexer which is used for keeping information about each word that occurs in an individual Web page. It consists of the three processes (Tokenizer, Token Filter and Aggregator) as shown in Figure (4.7), each will be explained below:

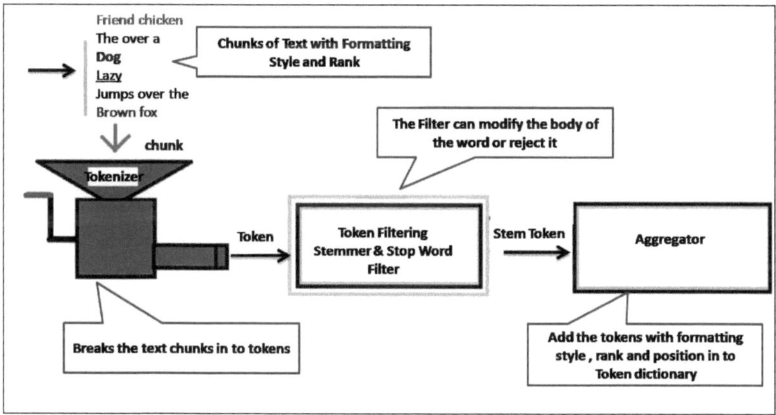

Figure (4.7) Text Tokenization

1. The Tokenizer

The Tokenizer is a component that splits the supplied text chunk (the text between two tags) into tokens. The input parameters of this process represent a chunk of text of the same style formatting and rank. It is not necessary for the bounds of the text chunks to match with the word boundary, allowing a word to span between two consecutive text chunks. An exemplification represents the below HTML markup:

```
<p>Lorem ipsum <a href="dolor-sit-amet">dolor sit amet</a>, consectetur
<strong>adipisicing elit<strong>, sed do eiusmod tempor incididunt ut labore et
<strong>dolore magna<strong> aliqua.</p>
```

The text chunks flow in the Tokenizer is explained in Figure (4.8) and the tokenizer algorithm is explained below:

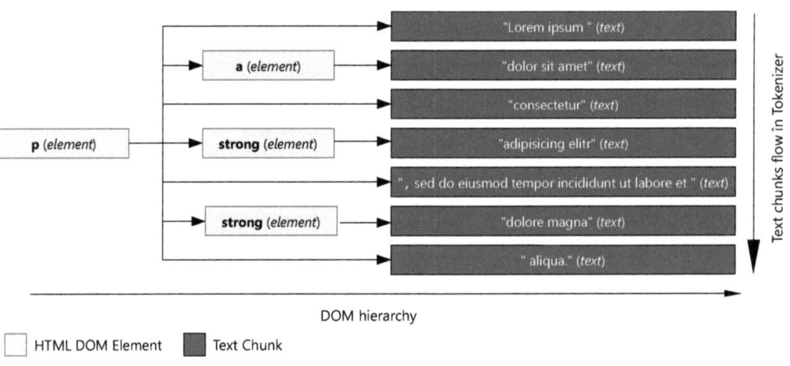

Figure (4.8) The flow of text chunks in the Tokenizer

Algorithm 4.3: The Tokenizer Algorithm	
Input parameters	
text	The text chunk to be tokenized.

style	The formatting style of this chunk.
rank	The rank level of this chunk.
Output Parameter	List of Tokens

```
Begin
     token ← incompleteToken
     forindex = 0 to length(text) - 1
        char ← text(index)
        if letter(char)  then token += char
        if isDigit(char) and length(token) = 0 then discard(char)
         else
            token += char
        endif
       if end_word_symbol (char) then
           move to next token
         aggregate(token, position, style, rank)
          position += 1
       endif
endfor
      if char(length(text)-1)<> end_word_symbol then
         incompleteToken← LastToken
     else
        incompleteToken← null
End.
```

2. Token Filtering

The Token Filtering can be performed by a separate module or by a custom Token Filtering implementation in the Search Engine Agent caller, since the Core Module exposes an interface to allow full control over this procedure.

Additionally, a built-in Token Filter called **Configurable Token Filter** supports defining the filtering process via XML-based configuration file. The filtering parameters specified in this configuration file consist of token length limits, non-alpha characters allowed in its body, stop-words, and stemming.

- **Adding Support for New Natural Languages**

While the scraping procedure requires a specific way to handle each particular natural language, the list of the available languages for scraping is limited to the current language support. To extend this support, new Token Filter Configuration files may be added to the “Filters” sub-directory from the application’s directory.

- **Stop Word Remover**

Stop word remover is the process of removing stop word term that has no meaning or irrelevant. Stop word remover is done by checking the stop word list, when the term matches one stop word list of contents, then the term is considered a stop word be discarded and will not be included on the stemming process. By the remover process will result terms without stop word.

The list of (1300) stop words in the system is used. This stop word list includes the articles (a, an, or the), prepositions (in, on, into, or at), pronouns (he, she, I, or me), and conjunctions (and, or, but, and so on) and the extracted or suggested stop words such as “repeat”, “high”, “speed”, “first”, etc. [31].

- **Stemmer:**

The stemmer can be defined by specifying a set of stemming rules executed by the built-in affix stemmer, either by pointing to an external custom stemmer. The custom stemmer can be any .NET Framework component which implements a stemmer interface exposed by the Core Module, thus allowing any stemming algorithm to drive the stemming procedure.

MWIM employs the Porter stemming algorithm to improve system accuracy by reducing the large number of words morphological variants. Terms with a common stem will usually have similar meanings, for example:
CONNECT, CONNECTED, CONNECTING, CONNECTION, CONNECTIONS
This may be done by removal of the various suffixes -ED, - ING, -ION; IONS to leave the single term CONNECT.

3. Aggregator

This process aggregates tokens in Content Extract's token dictionary along with its associated information (rank, formatting style, and position). To optimize the CE file compactness and memory usage, the formatting style information is stored without duplicates in a pool, from where it may be referenced by an individual token or a set of tokens.

4.5.1.3 The Reporting Module

The reporting module exports the information contained in CE binary file in a human-readable form, trough a stand-alone HTML document as shown in Figure (4.10).

The functionality of this module can be found in the "ProtoMWIM.Reporting.XHTML.dll" component library assembly file.

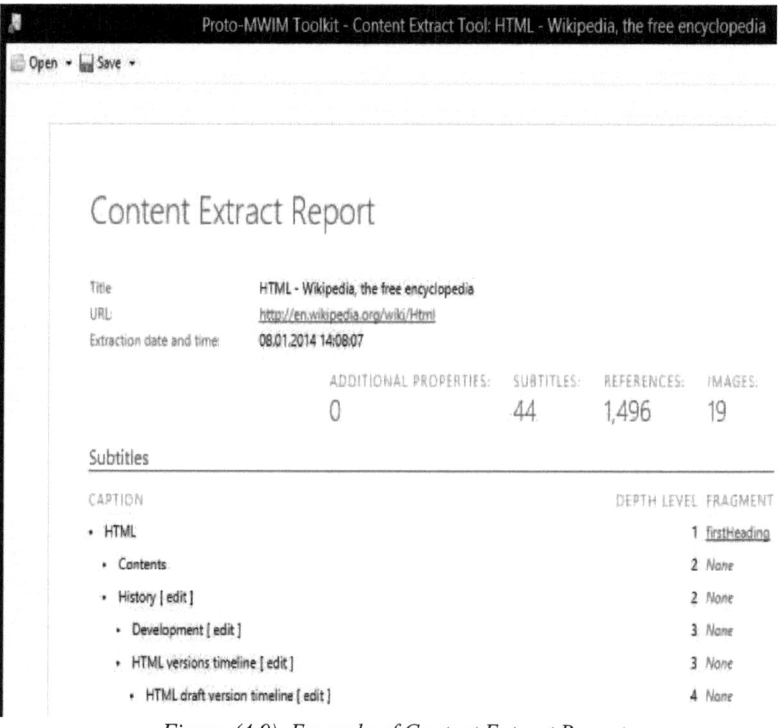

Figure (4.9) Example of Content Extract Report

4.5.2 The Proto-MWIM Toolkit

Modern computers ship with multi-core architectures, meaning that they have more than one processor. Generally managed applications do their work using only one processor. This makes things easier, but with this approach, you do not unleash all system resources. The reason is that all elaborations rely on a single processor that is overcharged and will take more time. Scaling applications across multiple processors is known as parallel computing, which is not something new in the programming world, in which the word "parallel" means that multiple tasks are executed concurrently, in parallel. The .NET Framework 4.0 provides support for parallel computing through the Task Parallel Library (also referred to as TPL).

The **Proto-MWIM Toolkit** is a software application which bundles a series of MWIM related tools for user to applied SEA functions. It uses a parallel computing for performing tasks concurrently, to take advantage of all logical CPUs available on a machine. This is improving the performance of this tool by reducing the time required for the finalization of the procedure.

CHAPTER FIVE

THE IMPLEMENTATION AND EVALUATION

5.1 Introduction

MWIM runs the functionality provided by the Search Engine Agent in two sides: Client side (Proto-MWIM Toolkit) and Server side (backend). This chapter describes the implementation of the functionality of the Search Engine Agent directly by using a graphical user interface tool set that called Proto-MWIM Toolkit without the need for any programmatic implementation.

Also, this chapter demonstrates an efficiency of a Proto- MWIM in two aspects (the network traffic savings and computational power required by the search engine servers) by presenting and discussing the results of applying Proto-MWIM Toolkit on a large set of Web pages.

5.2 The Proto-MWIM Toolkit Implementation

The Proto-MWIM Toolkit handles a series of MWIM related tools which allow the user to perform tasks such as:

- Generating a Content Extract metadata file by scraping a Web page;

- Inspecting the contents of a Content Extract file;
- Generating and updating automatically the Content Extract metadata files of a Website.

In order to run, this application requires .NET Framework 4 Client Profile installed on the target machine running a Windows XP with Service Pack 3 or a newer operating system.

5.2.1 Generating a Content Extract File by Scraping a Web Page

To scrape a Web page, it is necessary to launch the Content Extract Tool from the main window of the application.

After the window of this tool opens, by pressing the "Open" button from the toolbar, a menu provides two different opening options. In this case, the desired option will be "Web Page...". It will open the "Scrape Web Page" dialog, displayed in the Figure (5.1) below, which contains the scraping options.

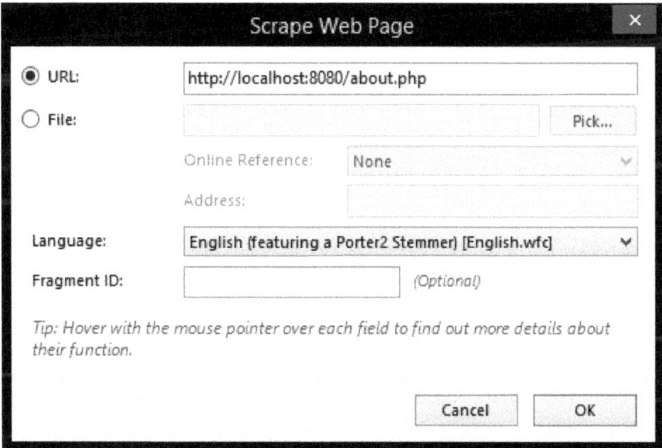

Figure (5.1) The Web Page Scraping Options Dialog

- **"Url" Radio Button**

The "Url" radio button should be checked, if the target Web page is hosted by a Web server, or needs to be pre-processed by a local server, The URL of the document should be specified in the text box located in the right side of this radio button.

- **"File" Radio Button**

The "File" checkbox should be checked if the Web page represents a HTML file located on the local storage. The file can be selected by pressing the "Pick" button.

In the case of the local HTML files, the URL of the existing online copy of the document (if the computer is offline) can be attributed to the resulted Content Extract by selecting "Web Page's URL" on the "Online Reference" field and specifying the URL address in the "Address" field. In the same case, if the URL of the document is unknown, but the base URL for resolving the relative references is available, it can be specified in a similar way, but selecting "Base URL" for the "Online Reference" field. This step is not necessary if the HTML document has the base URL specified in the header.

***Note**: An URL address specified in an address field must include the scheme identifier ("http://").*

- **"Language" field**

The natural language in which the document is written can be selected from the "Language" field. It's not necessary to specify the language for a correct tokenization and pre-processing (stop-word filtering and stemming) of the contents. To disable the pre-processing and using the default tokenization rules, "(None)" can be selected as the value of this field.

- **"Fragment ID" field**

If the scraping is desired to be limited to a certain fragment of the document, the HTML ID of the desired fragment can be specified on the "Fragment ID" field. If the specified fragment cannot be found in the document, it will be ignored and the scraping procedure will be performed to the entire content of the document.

The scraping procedure begins after the "OK" button is pressed. If the procedure succeeds, its Content Extract report will be displayed on the "Content Extract Tool" window. The Content Extract file can be saved by pressing the "Save" button found in the toolbar, then selecting the "Content Extract file".

5.2.2 Inspecting the Contents of a Content Extract

The user can inspect a report about the information stored in a Content Extract file using the Content Extract Tool, by opening a Content Extract file located on the storage drives of the computer which runs the application. This can be done by pressing the "Open" button from the toolbar, then selecting the "Content Extract File..." menu item.

The Content Extract report lists all information contained by it, excepting the token positions which are not too relevant for a human being and for compactness reasons.

The user can save the report as a single-file XHTML document by pressing the "Save" button from the toolbar, then selecting the "Report..." menu item. The report displayed by this tool is generated by the Proto-MWIM XHTML Reporting Module.

5.2.3 Generating and Updating Automatically the Metadata of a Website.

In the case of static Websites, or Websites in which the back-end technologies are not able to interoperate with Proto-MWIM SDK, their metadata can be generated automatically by using the Website Scraper tool. This tool can automatically infer the structure of a Website for generating the Sitemap, and generates the Content Extract file for each Web page, according to the user's settings.

After launching the Website Tool, the "Website Profiles" are available for generating and updating the metadata.

A **Website profile** contains the scraping options particular to each Website. A new Website profile can be created by pressing the "Add new..." button found in the toolbar of the window. A profile editing dialog will appear, as shown in the Figure (5.2), where the settings of the Website profile must be specified. Each Website profile must have a unique name, which will be displayed in the Website profiles window. It should be specified in the "Profile Name" field. In order to store the metadata along with the Website files, the local directory which is the root of the Website should be selected by pressing the "Pick" from the right side of the "Root directory (local)" field. The URL address of the Website should also be specified in the "URL" field, to generate correctly the Sitemap information and the Content Extract. It is not required for this address to be accessible during the Website scraping procedure. If the Website is a dynamic one, the Web pages can be retrieved through a local host server. To enable this feature, the "Retrieve pages through the local host" checkbox should be checked. Also, the URL address of the Website's root on the local host should be filled in the field located on the right side of the same checkbox.

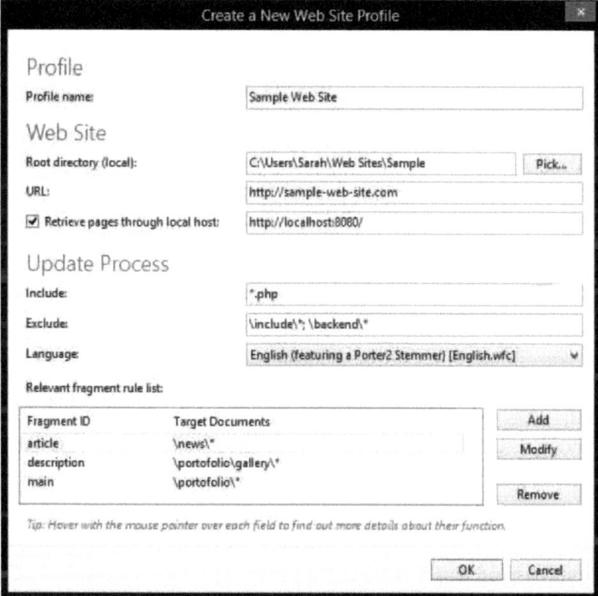

(Figure 5.2) The Profile Editing Window of the Website Scraper Tool

The structure of the Website is established in two steps:

1. Using the information from the Sitemap;
2. By discovering new pages not recorded in the Sitemap by scanning the Website's root directory and sub-directories for locating the source files of the Web pages (.htm, .html, .xht and .xhtml).

The type and location of the files which will be considered Web pages, and will be seeking for in the second stage, can be specified as wildcard patterns in the "Include" field. Details about the supported wildcard queries can be found later in this section.

For default, the file patterns point to the HTML files located in the root directory of the Website and its sub-directories. This is not appropriate in the case of dynamic Websites, where the source files for pages can be of different types.

Thus a Website built on PHP server software may need to add the .php file extension in the pattern from the "Include" field. The wildcard query used to locate the Web page source documents in a PHP-based Website would be:

```
*.htm; *.html; *.xht; *.xhtml; *.php
```

Now if a .php file is found in the root directory of a Website or in its sub-directories, the file will be retrieved through local host server (assuming that this feature is enabled) scraping its output and adding a record for it in Sitemap.

To avoid considering unwanted particular files as being Web pages, their wildcard query can be specified to the "Exclude" field to blacklist all files matching with this query.

For example, to exclude the .php files located in the "include" folder, which do not represent a Web page, but a shared codebase between the Web pages, a The wildcard query like the one below can be used for the "Exclude" field to ignore all ".php" files from the "include" directory and its sub-directories.

```
*\include\*.php
```

To completely disable the second step in automatically inferring the Website's structure, it is necessary to blacklist all files by setting on the "Exclude" field. The wildcard query "*"used to exclude all files, disabling the automatic Website structure discovery feature.

In the case where a page source file may generate a different output according to the query segment of the URL, it is necessary that the query string corresponding to each different output variant must be manually specified as a separate record in the Sitemap. Otherwise, these variants will not be scrapped, because the Website structure discovery algorithm is not aware of them.

If the scraping is desired to be limited to a certain fragment of a document, a rule list can define a set of fragment identifiers with their corresponding documents specified as a wildcard character.

This list is located on the bottom of the dialog. A new rule can be added by pressing the "Add" button from the right side of this list. Later, a rule may be edited or removed by using the "Modify" and "Remove" button located on the same side of the list.

Back to the profiles window, a profile may be edited again by pressing the "Edit..." button, or removed by pressing the "Remove" button, both located in the window toolbar. The profiles are saved automatically.

By pressing the "Update" button, the metadata of the Website will be generated or updated. This procedure will generate a Content Extract for each new Web page (except on the initial update, where this behavior applies to each page) and updates the modified ones, and then updates the Sitemap accordingly.

After the procedure completes, the details about the procedure are logged in the activity log, which can be opened by pressing the "Show activity log" link found on the main window of this application. The log can be cleared from the "Options" dialog, which can be launched from the main window.

The date and time of the last update of each Website is displayed in the profile list. The removal of all MWIM-related metadata of a Website can be done by pressing the "Clear" button from the Website profiles window's toolbar. This will delete all Content Extract files and their records from Sitemap. The Sitemap records which describe the Website structure will not be removed.

- **Supported Wildcard Patterns**

A wildcard pattern is a series of characters that are matched against incoming character strings. As seen previously, some fields used in this application requires wildcard patterns to match file paths. The wildcard characters supported are listed in the table (5.1).

Table (5.1) The list of the supported wildcard patterns

Wildcard Character	Description
*	Matches one or more characters.
?	Matches one character.
#	Matches one digit.

Additionally, a wildcard character may be followed by a range of characters enclosed in square brackets, which represents the match condition. The supported range syntaxes are listed and exemplified in the table (5.2).

Table (5.2) The list of the supported wildcard range conditions

Range Syntax	Matches
[List]	Any character included inside of the brackets. **Example:**?[aeiou] matches one vowel.
[Start]-[End]	Any characters inside of the alphabetic range of two characters. **Example:**?[a-m] matches any character alphabetically located between “a” and “m”.
[!List]	Any other characters but not these specified

	inside of the brackets. **Example:**?[!aeiou] matches one consonant.

Multiple wildcard patterns may be specified for a query delimited by a semicolon character (;).

The pattern matching is performed on the path and file name of a file, relative to the root directory of a Website. It begins with the "\" to allow matching the root directory when is needed. For example, the following query will match all files from the "misc" directory and its sub-directories located on the root of the document:

```
\misc\*.*
```

While this matches all files from a directory and its sub-directories, located anywhere in the Website's sub-directory branch, but not the one located on the root directory:

```
*\misc\*.*
```

The path of the "misc" sub-directory, child of the root directory is "\misc", thus the above query will not match this sub-directory, because there is no string available behind of the backslash leading the directory name, as this pattern is expecting.

5.3 MWIM Experimentation and Evaluation

MWIM experiment consists of collecting statistics report resulted from scraping different HTML document samples using Proto-MWIM Toolkit.

Exclusively for this experiment, a feature has been implemented in Proto-MWIM Toolkit to collect details about each HTML document and its processing details, logging them to a CSV data table file referred from now as **Statistics Table**. Each record from the Statistics Table contains the following fields:

- **URL** - The URL address of the document of which the current record belongs. The URL address is relative to the Website's root.
- **Original Size** - The size of the document.
- **CE size** - The size of the resulted Content Extract file.
- **Parsing Time** - The time spent for parsing the HTML markup code and generating its DOM representation.
- **Processing Time** - The time spent for retrieving and pre-processing the information stored in the Content Extract file.
- **Overall Time** - The time spent by Proto-MWIM Toolkit to complete the job. It includes additional time spending aside of scraping the document and serializing the CE, such as generating the hash of the document, retrieving the Content Extract deserialization time, and from other logics that manage the procedure.
- **CE serialization Time** - The time spent for saving the CE information to a file.
- **CE Deserialization Time** - The time spent for retrieving the CE information from a file.

The value of the fields which describes file sizes are using the byte measure unit, while the ones which describe a time, use the milliseconds as the measuring unit.

The statistics report generating features could be disabled for default. It can be enabled by checking a "log statistics about Website scraper's jobs" that found in options of Proto-MWIM Toolkit. Figure (5.3) shows a part from the statistics report.

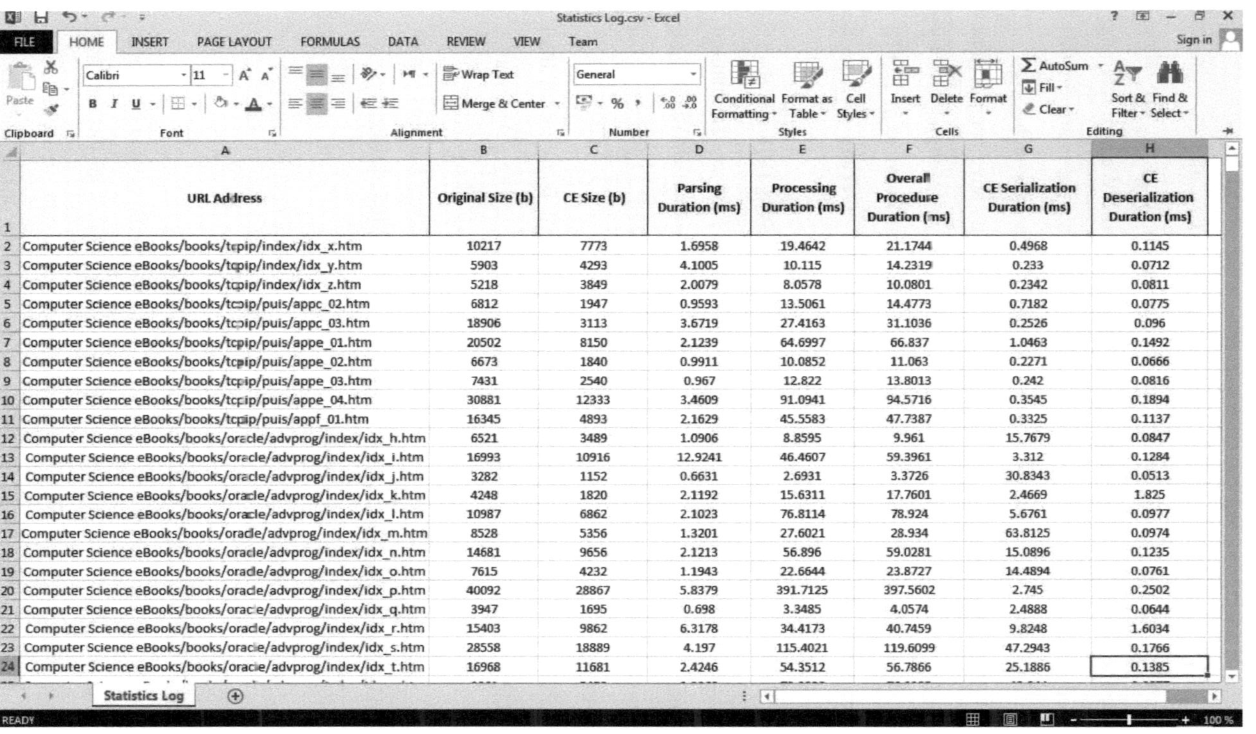

URL Address	Original Size (b)	CE Size (b)	Parsing Duration (ms)	Processing Duration (ms)	Overall Procedure Duration (ms)	CE Serialization Duration (ms)	CE Deserialization Duration (ms)
Computer Science eBooks/books/tcpip/index/idx_x.htm	10217	7773	1.6958	19.4642	21.1744	0.4968	0.1145
Computer Science eBooks/books/tcpip/index/idx_y.htm	5903	4293	4.1005	10.115	14.2319	0.233	0.0712
Computer Science eBooks/books/tcpip/index/idx_z.htm	5218	3849	2.0079	8.0578	10.0801	0.2342	0.0811
Computer Science eBooks/books/tcpip/puis/appc_02.htm	6812	1947	0.9593	13.5061	14.4773	0.7182	0.0775
Computer Science eBooks/books/tcpip/puis/appc_03.htm	18906	3113	3.6719	27.4163	31.1036	0.2526	0.096
Computer Science eBooks/books/tcpip/puis/appe_01.htm	20502	8150	2.1239	64.6997	66.837	1.0463	0.1492
Computer Science eBooks/books/tcpip/puis/appe_02.htm	6673	1840	0.9911	10.0852	11.063	0.2271	0.0666
Computer Science eBooks/books/tcpip/puis/appe_03.htm	7431	2540	0.967	12.822	13.8013	0.242	0.0816
Computer Science eBooks/books/tcpip/puis/appe_04.htm	30881	12333	3.4609	91.0941	94.5716	0.3545	0.1894
Computer Science eBooks/books/tcpip/puis/appf_01.htm	16345	4893	2.1629	45.5583	47.7387	0.3325	0.1137
Computer Science eBooks/books/oracle/advprog/index/idx_h.htm	6521	3489	1.0906	8.8595	9.961	15.7679	0.0847
Computer Science eBooks/books/oracle/advprog/index/idx_i.htm	16993	10916	12.9241	46.4607	59.3961	3.312	0.1284
Computer Science eBooks/books/oracle/advprog/index/idx_j.htm	3282	1152	0.6631	2.6931	3.3726	30.8343	0.0513
Computer Science eBooks/books/oracle/advprog/index/idx_k.htm	4248	1820	2.1192	15.6311	17.7601	2.4669	1.825
Computer Science eBooks/books/oracle/advprog/index/idx_l.htm	10987	6862	2.1023	76.8114	78.924	5.6761	0.0977
Computer Science eBooks/books/oracle/advprog/index/idx_m.htm	8528	5356	1.3201	27.6021	28.934	63.8125	0.0974
Computer Science eBooks/books/oracle/advprog/index/idx_n.htm	14681	9656	2.1213	56.896	59.0281	15.0896	0.1235
Computer Science eBooks/books/oracle/advprog/index/idx_o.htm	7615	4232	1.1943	22.6644	23.8727	14.4894	0.0761
Computer Science eBooks/books/oracle/advprog/index/idx_p.htm	40092	28867	5.8379	391.7125	397.5602	2.745	0.2502
Computer Science eBooks/books/oracle/advprog/index/idx_q.htm	3947	1695	0.698	3.3485	4.0574	2.4888	0.0644
Computer Science eBooks/books/oracle/advprog/index/idx_r.htm	15403	9862	6.3178	34.4173	40.7459	9.8248	1.6034
Computer Science eBooks/books/oracle/advprog/index/idx_s.htm	28558	18889	4.197	115.4021	119.6099	47.2943	0.1766
Computer Science eBooks/books/oracle/advprog/index/idx_t.htm	16968	11681	2.4246	54.3512	56.7866	25.1886	0.1385

Figure (5.3) A part of Statistics Report

The resulted Statistics Table file is located in the "Statistics" directory, which is a sub-directory of the application directory. The name of the file will be specified in the Activity Log, after the Website scraping procedure completes. These statistics details will be used later in this section to support the claimed benefits brought by this approach.

A relatively large number of HTML documents have been used as a sample for this experiment, as the accuracy of the results is directly proportional to the number of testing documents. **The count** is of **22,678** documents with a **total size** of **357.8** MB. The **average size** of a document has been around of **16.1** KB.

These documents are converged together to build a fictive Website, to be scraped later using the Proto-MWIM Website Scraper. The natural language used for the content of these documents is English, the default token filter being set for this operation. It consists of a stop word list counting 1300 items and a Porter stemmer.

The operation has been performed on a system featuring an Intel Core 2 Duo processor with a clock rate of 2 GHz, 2 GB RAM with a clock rate of 800 MHz. It completes in 16 minutes and 16 seconds.

This test will focus on evidence two of the key benefits which contribute on a large scale to the overall efficiency of the concept:

1. The network traffic savings.
2. Less computational power required by the search engine servers.

5.3.1 Determining the Network Traffic Savings

One of the factors that reduce the network traffic caused by search engines consists of retrieving only the information relevant for Web indexing purposes, instead of the whole markup code of a page. The amount of savings will be established by deterring the difference between the size of the markup code and the size of the Content Extract metadata file for the test Website.

Considering S_M as the sum of the "Original Size" fields of all records from the Statistics Table resulted from this experiment, respectively S_C as the sum of the "CE Size" fields of the same records, the amount of savings will be: $\Delta_S=S_M-S_C$.

The values of SM and SC have been shown to be of **375,254,713** bytes (**357.8 MB**) respectively **101,605,594** bytes (**96.8** MB). According to these values, Δ_S becomes **237,649,119** (**261** MB).
Thus, for a full indexing of the test Website, the search engine would have retrieved **261** MB less in the MWIM approach than using the traditional method.

The average size of a Content Extract is 27.1% of the original HTML markup size (the average size of a Content Extract metadata file being of one-third of the original HTML markup size), thus the percentage of savings resulted to be of **72.9%**.

5.3.2 Establishing the Time Resources Spent by Using MWIM

This section aims to establish the average time spent in different stages of the MWIM forward indexing according to the results obtained from this experiment.
The Web indexing tasks that would be performed on the regular Web hosting servers, where split in three major stages:

- **Document Parsing** - The stage where a DOM tree representation is retrieved from a (X)HTML markup code.
- **Information Collection** - The stage where the information required in a Content Extract is extracted from the DOM tree representation resulted from the previous stage. This stage includes the natural language processing.

- **Content Extract Serialization** - In this stage the information extracted from the previous stage is serialized to a binary file.

The average time spent for the Document Parsing Stage (**T_P**), Information Collection Stage (**T_{IC}**), and Content Extract Serialization Stage (**T_S**) have been established by determining the average value of the "Parsing Time", "Processing Time" respectively "CE Serialization Time" fields of all Statistics Table records.

Their values are shown to be of: **7.12** ms for **T_P**, **237.78** ms for **T_{IC}**, respectively **21.18** ms for **T_S**. The average time spent for all three stages was of **266.08** ms per document.

On the search engine servers, the **average time** spent in the Content Extract Deserialization stage (**T_D**) was **0.33** ms, being established by calculating the average value of the "CE Deserialization" field from all records of the Statistics Table.

An overview of the timings of all stages is illustrated in Figure (5.4).

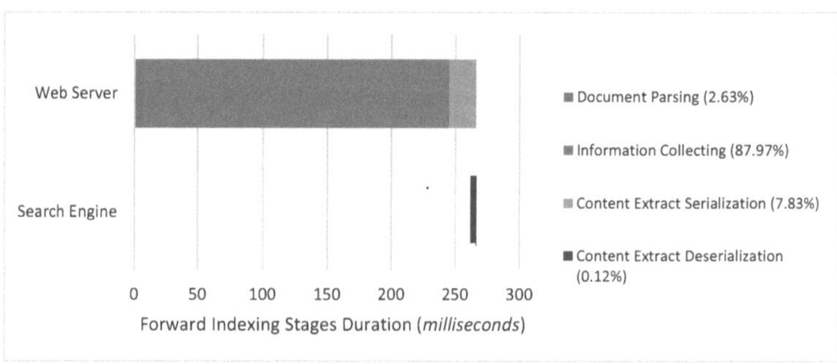

Figure (5.4) The Timeline of The MWIM Forward Indexing Stages, As Resulted in This Experiment.

As can be observed in the Figure above, a large extent of the processing time involved in forward indexing has been shifted from the search engine severs to regular Web servers.

The parsing stage and the natural language processing stage which is performed on the search engine servers in the traditional Web indexing method are replaced in the MWIM approach with only one stage: Content Extract deserialization. This stage according to this experiment has a similar duration like the parsing stage, eliminating the time spending on the search engine servers for the natural language processing tasks.

Thus, the search engine servers require less computational power to retrieve the forward index of a Web page, more becoming available for indexing Websites by using the traditional methods.

CHAPTER SIX

CONCLUSIONS AND FUTURE WORKS

6.1 Conclusions:

The MWIM concept brings a series of advantages and disadvantages, as follows:

- **Less Network Bandwidth Consumption** - unlike the standard search engine which retrieves each Web page of a Website, in this case, are retrieved the CE files only for the modified or newly created Web pages. Based on a binary format optimized for size compactness and containing only the data extraction results, the CE files are with 72.9% smaller in size than the original HTML code according to the efficiency experiment from the previous chapter. Due to these two aspects, drastic traffic savings on the World Wide Web network infrastructure are made.
- **Less Computing Power Consumption in Search Engines** – As the Web indexing computational requirements are partially distributed on the hardware of the Web hosting servers, the computational power required by Search Engines for indexing the MWIM enabled Websites is significantly reduced. As observed in the Proto-MWIM efficiency experiment, the search engines would

spend significantly less time deserializing a CE file than parsing the original document.

- **Faster Reflection of Content Modification in Search Results** - due to reassigning the parsing and data extraction to the Website back ends, the routine of the search engine data extraction procedure resumes only in retrieving and aggregating the pre-processed (extracted) data. This causes faster data updating cycles for the index in search engines and the addition or modification of Web pages, would be reflected faster in the search results. This would represent a benefit for the Websites with highly dynamic content such as the news providers.
- **More Accurate Search Results** – Due to the fact of the content is submitted selectively to the search engines, it can be filtered by the Websites of any irrelevant portions such as the header, navigation menu, footer and related articles suggestion box from a Web page, keeping only the portions which relates to the main subject.
- **Reducing Crawler and indexer Time-** Sites using Sitemaps tend to be crawled and re-crawled faster. A sitemap helps the crawler and the indexer to know about the location of the pages site, CE files, and determined the last update in any page on the Website.
- **Additional Storage Space Consumption** - An additional storage space is required on the Website's hosting server in order to store the metadata, whose size depends of the Website's contents. As resulted from the efficiency experiment, the average size of a Content Extract is 27.1% of the original HTML markup size. Thus a Proto-MWIM-enabled Web site would require at least 27.1% more space than a traditional Web site.

- **Additional Computational Power Consumption in Web Servers** As the Web indexing tasks were partially shifted from the search engine server's machine to the machine of the Website's hosting server, the parsing of a Web page or document and its data processing will be executed on the CPU(s) of the machine which provides the hosting of the Website. However, this aspect may be negligible since these tasks are executed only when the contents of the Website got changed.
- **Explicit Implementation** - In order to be applicable, the MWIM approach needs to be explicitly implemented in the Web site's back end software by the developer. Thus, an additional effort is required to promote the Website as being a MWIM enabled one.

6.2 Suggestions for Future Works

The following advantages are a matter of future work, currently being not detailed in this paper:

- **Up-to-date index updating by using Multilateral Web Indexing Client Dispatcher (MWICD)**- For MWIM approach to work more efficiently, it is necessary to establish a service called Multilateral Web Indexing Client Dispatcher, abbreviated as MWICD. This service is suggested to establish an up-to-date indexing updating process in search engine where the Website metadata is up-to-date updating immediately after each modification which provides a list of all MWIM enabled Websites and broadcasting the content modification notifications from the Search Engine Agents to the search engines servers. Once the search engine receives a content modification notification, it retrieves the sitemap of that Website and compares it to the cached version to determine which Web pages, documents or multimedia resources have been added, modified or removed in order to re-index these elements by retrieving and further processing the data contained in the Content Extract files. Another function of MWICD is to host Search Engine Agent updates, which need to auto-update in order to quickly deliver security updates or eventual new futures. Since the search engine is notified through the broadcasts from MWICD about the modifications incurred to a Website, the indexing can be performed in a relatively short time, thus improving significantly the time required for these changes to reflect them in the search results.

- **A Broader Range of Source File Formats and languages** - Generated metadata for a broad variety of source file formats such as (.pdf, .doc, .ppt, .xsl, etc.) and multimedia resources contained by each Web page. Also, for other languages such as Arabic Language.

References:

[1] Liu B., **Web Data Mining**, Springer-Verlag, 2007.

[2] Fadhil M. N., **A Proposed System for Semantic-Based Internet Search Engine**, MSc. Thesis, Computer Science Department, University of technology, Iraq, September – 2004.

[3] Estiévenart F., François A., Henrard J. and Hainaut J., **A Tool-Supported Method to Extract Data And Schema From Web Sites**, Website evolution, Fifth IEEE international workshop, 2003.

[4] LAN YI , **Web Page Cleaning for Web Mining,** Ph.d Thesis, Computer Science Department, National University of Singapore, 2004.

[5] Migletz J., **Automated Metadata Extraction,** MSc. thesis, Naval Postgraduate School, June 2008.

[6] Al-Attar I. T., **Design and Implementation of Web Search Engine,** Ph.d Thesis, Computer Science Department, University of Technology, Iraq, June-2009.

[7] Choochaiwattana W., **An Algorithm of Product Information Extraction from Web Pages: a Document Object Model Analysis Approach,** 2nd International Conference on Information Communication and Management (ICICM2012), pages 103 -107, Singapore, 2012.

[8] Yang J. and et al, **Incorporating Site-Level Knowledge to Extract Structured Data from Web Forums**, ACM, Madrid, Spain, April 20–24, 2009.

[9] Rana R. K., Tyagi N., **A Novel Architecture of Ontology-based Semantic Web Crawler**, International Journal of Computer Applications (0975 – 8887), Volume 44– No18, April 2012.

[10] Behrouz A. Forouzan, **Data Communications and Networking,** Fourth Edition, McGraw-Hil, 2007.

[11] Naik U., Shivalingaiah D., **Comparative Study of Web 1.0, Web 2.0 and Web 3.0,** 6th International CALIBER -2008, University of Allahabad, February 28-29 & March 1, 2008.

[12] Aghaei S., Nematbakhs M. A., Farsani H. K., **Evolution of The World Wide Web: From Web 1.0 to Web 4.0,** International Journal of Web & Semantic Technology (IJWesT), Vol.3, No.1, January 2012.

[13] Al-Umary E. S., **A Proposed For Enhanced E-Commerce Searcher**, MSc. Thesis, Computer Science Department, University of Technology, Iraq, February - 2006.

[14] Velásquez J., Palade V., **Adaptive Web Sites,** IOS Press, 2008.

[15] Baca M., **Introduction to Metadata,** Second Edition, J. Paul Getty Trust, 2008.

[16] Gourley D., Totty B., Sayer M., Reddy S. and Aggarwal A., **HTTP the Definitive Guide,** O'Reilly Media, 2002.

[17] Huddleston R., **HTML, XHTML, and CSS: Your visual blueprint™ for designing effective Web pages,** Wiley Publishing, Inc., Indianapolis, Indiana, 2008.

[18] Baldi P., Frasconi P. and Smyth P., **Modeling the Internet and the Web: Probabilistic Methods and Algorithms**, John Wiley & Sons Ltd, 2003.

[19] Vrieze P., **Improving search engine technology**, MSc. Thesis, Paul de Vrieze, Mar- 2002.

[20] Farrel A., **The Internet and Its Protocols A Comparative Approach**, Morgan Kaufmann, 2004.

[21] Lee T. B., **Uniform Resource Identifier (URI): Generic Syntax,** RFC 3986, the Internet Society, January 2005.

[22] Snell N., Temple B., Clark T., **Sams Teach Yourself Internet and Web Basics All in One**, Sams, 2003.

[23] Thomas S., **HTTP Essentials Protocols for Secure, Scaleable Web Sites Stephen,** John Wiley & Sons, Inc., 2001.

[24] Chakrabarti S., **Mining the Web: Discovering Knowledge from Hypertext Data**, Morgan Kaufmann, 2003.

[25] Musciano CH., Kennedy B., **HTML & XHTML: The Definitive** Guide, 6th Edition, O'Reilly & Associates, Inc, 2008.

[26] Jerkovic J. I., **SEO Warrior**, O'Reilly Media, Inc.,2010.

[27] Skibinsk P., **Improving HTML Compression**. Data Compression Conference, DCC 2008 , Page(s): 545, 2008.

[28] Loioyd I., **The_Ultimate_HTML_Reference**, SitePoint Pty Ltd,. 2008.

[29] Eckstein R., Spainhour S., **Webmaster in a Nutshell, 3rd Edition**, O'Reilly Media, 2002.

[30] Duckett J.,**Wiley**, **Beginning HTML, XHTML, CSS, and JavaScript,** Wiley Publishing, Inc, 2010.

[31] Hashim S. M., **Classification Of HTML and PDF Documents By Using Association Rules**, MSc. Thesis, Commission for Computers and Informatics, Iraq, February – 2012.

[32] Carey P., **New Perspectives on HTML, XHTML, and XML,** 3rd Edition, Course Technology, Cengage Learning, 2010.

[33] Melton J.,Buxton S., **Querying XML XQuery, XPath, and SQL/XML**, Morgan Kaufmann, 2006.

[34] Skonnard A.,Gudgin M., **Essential XML Quick Reference**, Addison wisely, 2002.

[35] Marini J., **The Document Object Model, Processing Structured Documents**, McGraw-Hill/Osborne, 2002.

[36] Tennison J., **XSLT and XPath on the Edge**, Hungry Minds, Inc., 2001.

[37] Watt A., **XPath Essentials**, John Wiley & Sons, Inc., 2002.

[38] Zou J., Le D., Thoma G.R., **Combining DOM Tree and Geometric Layout Analysis for Online Medical Journal Article Segmentation,** ACM, Chapel Hill, North Carolina, USA, 2006.

[39] Pappas P., Katsimpras G., Stamatatos E., **Extracting Informative Textual Parts from Web Pages Containing User-Generated Content**, the 12th International Conference on Knowledge Management and Knowledge Technologies, ACM, New York, NY, USA, 2012.

[40] Lin SH., Chu K., Chiu CH., **Automatic sitemaps generation: Exploring website structures using block extraction and hyperlink analysis**, Expert Systems with Applications,38, pages 3944–3958,Elsevier Ltd, 2011.

[41] Olfat H., **Automatic Spatial Metadata Updating and Enrichment,** PH.D Thesis , Department of Infrastructure Engineering, School of Engineering, The University of Melbourne, Victoria, Australia, January-2013.

[42] Litwin L., Ross M., **Geoinformation Metadata in INSPIRE and SDI: Understanding. Editing. Publishing**, Springer, 2011.

[43] Urbani J., **RDFS/OWL reasoning using the MapReduce framework**, MSc. Thesis, Department of Computer Science, Vrije University, Amsterdam, July, 2009.

[44] Daconta M. C., Obrst L. J., Smith K. T., **The Semantic Web: A Guide to the Future of XML, Web Services, and Knowledge Managemen**, Wiley Publishing, Inc, 2003.

[45] Candan K. S., Liu H., Suvarna R., **Resource Description Framework:Metadata and Its Applications**, ACM SIGKDD Explorations Newsletter, Volume 3 Issue 1, Pages 6-19, July 2001.

[46] Naeem M., Bashi O. T., **RDFa as Semantic Markup and Web Visibility**, M.Sc. Thesis, School of Technology, Malmo University spring 2011.

[47] Antoniou G., Harmelen F., **A Semantic Web Primer,** 2nd Edition, the MIT Press Cambridge, Massachusetts, England, 2008.

[48] Bruijn J., Fensel D., Kerrigan K., Keller U., Lausen H·, Scicluna J., **Modeling Semantic Web Services**, Springer-Verlag Berlin Heidelberg, 2008.

[49] Duval E., Hodgins W., Sutton S., Weibel S. L., **Metadata Principles and Practicalities**, D-Lib Magazine, Vol 8 Number 4, 2002.

[50] Ferrara E., Demeo P., Fiumara G., Baumgartner R., **Web Data Extraction, Applications and Techniques: A Survey,** ACM Computing Surveys, Pages 1-54, July 2012.

[51] Chang CH., Kayed M., Girgis M.R., Shaalan K., **A Survey of Web Information Extraction Systems**, IEEE Transactions on Knowledge and Data Engineering, 2006.

[52] Devika K., Surendran S., **An Overview of Web Data Extraction Techniques**, International Journal of Scientific Engineering and Technology, Volume 2 Issue 4, pp : 278-287 , April 2013.

[53] Das A., Nandy S., **Hybrid Focused Crawler - A Fast Retrival of Topic Related Web Resource For Domain Specific Searching,**

International Journal of Information Technology and Knowledge Management, Vol. 2, No. 2, pp. 355-360, July-December 2010.

[54] Swami SH. A., Vidap P., **Web Data Extraction and Alignment Tools: A Survey,** International Journal of Information Technology and Knowledge Management, Volume No.2, Issue No.6, pp : 573-578, June -2013.

[55] Gupta S., Bhatia K., Manchanda P., **WebParF:A Web Partitioning Framework for Parallel Crawler,** International Journal on Computer Science and Engineering (IJCSE), Vol. 5, No. 08, Aug - 2013.

[56] Kumar M. S., Neelima P., **Design and Implementation of Scalable, Fully Distributed Web Crawler for a Web Search Engine,** International Journal of Computer Applications, Vol. 15, No.7, February 2011.

[57] Levene M., **An Introduction to Search Engines and Web Navigation,** John Wiley & Sons, Inc, 2010.

[58] Khurana D., Kumar D., **Web Crawler: A Review**, International Journal of Computer Science & Management Studies(IJCSMS), Vol. 12, Issue 01, January- 2012.

[59] Manning CH. D., Raghavan P., Schütze H., **An Introduction to Information Retrieval,** Cambridge University Press, 2009.

[60] Felix Van de Maele **,Ontology-based Crawler for the Semantic Web,** MSc. Thesis, Department of Applied Computer Science, Vrije Universitiet Brussel, May-2006.

[61] SPINK A., JANSEN B. j., **Web Search : Public Searching of The Web**, Kluwer Academic Publishers, 2004.

[62] Peshave M., **How Search Engines Work and A Web Crawler Application,** CiteSeerX, 2005.

[63] Rajan R. H., Dhas J., **A Method for Classification based on Association Rules using Ontology in Web Data,** International Journal of Computer Applications , Vol. 49,No.8, July- 2012.

[64] Giridhar N. S, Prema K.V., Reddy N. V. S., **A Prospective Study of Stemming Algorithms for Web Text Mining,** Ganpat University /Journal of Engineering and Technology, Vol.-1, Issue-1, Jan-Jun-2011.

[65] M.F.Porter, **An algorithm for suffix stripping**, Program, Vol.14, No.3, pp 130-137, July-1980.

[66] Mishra A. A., Kamat CH., **Migration of Search Engine Process into the Cloud,** International Journal of Computer Applications, Vol.19, No.1, April- 2011.

[67] Ah Kioon M. C., Wang Z. and Deb Das SH., **Security Analysis of MD5 algorithm in Password Storage**, the 2nd International Symposium on Computer, Communication, Control and Automation (ISCCCA-13), Atlantis Press, Paris, France, 2013.

Contents

Chapter Four: Multilateral Web Indexing Model

Chapter Five: The Implementation and Evaluation of MWIM

Chapter Six: Conclusions and Future Works

Printed by Books on Demand GmbH, Norderstedt / Germany